SCIENCE POLICIES AND TWENTIETH-CENTURY DICTATORSHIPS

Science, Technology and Culture, 1700–1945

Series Editors

Robert M. Brain
The University of British Columbia, Canada

and

Ernst Hamm
York University, Canada

Science, Technology and Culture, 1700–1945 focuses on the social, cultural, industrial and economic contexts of science and technology from the 'scientific revolution' up to the Second World War. Publishing lively, original, innovative research across a broad spectrum of subjects and genres by an international list of authors, the series has a global compass that concerns the development of modern science in all regions of the world. Subjects may range from close studies of particular sciences and problems to cultural and social histories of science, technology and biomedicine; accounts of scientific travel and exploration; transnational histories of scientific and technological change; monographs examining instruments, their makers and users; the material and visual cultures of science; contextual studies of institutions and of individual scientists, engineers and popularizers of science; and well-edited volumes of essays on themes in the field.

Also in the series

*Acid Rain and the Rise of the Environmental Chemist
in Nineteenth-Century Britain*
Peter Reed

*Matter and Method in the Long Chemical Revolution
Laws of Another Order*
Victor D. Boantza

*From Local Patriotism to a Planetary Perspective
Impact Crater Research in Germany, 1930s–1970s*
Martina Kölbl-Ebert

Science Policies and Twentieth-Century Dictatorships
Spain, Italy and Argentina

Edited by

AMPARO GÓMEZ
Universidad de La Laguna, Spain

ANTONIO FCO. CANALES
Universidad de La Laguna, Spain

BRIAN BALMER
University College London, UK

Routledge
Taylor & Francis Group

LONDON AND NEW YORK

First published 2015 by Ashgate Publishing

2 Park Square, Milton Park, Abingdon, Oxfordshire OX14 4RN
52 Vanderbilt Avenue, New York, NY 10017

Routledge is an imprint of the Taylor & Francis Group, an informa business

First issued in paperback 2020

British Library Cataloguing in Publication Data
A catalogue record for this book is available from the British Library

The Library of Congress has cataloged the printed edition as follows:
Science Policies and Twentieth-Century Dictatorships: Spain, Italy and Argentina / edited
    by Amparo Gómez, Antonio Fco. Canales and Brian Balmer.
       pages   cm. – (Science, Technology and Culture, 1700–1945)
    Includes bibliographical references and index.
    1. Science and state – Spain – History – 20th century. 2. Science and state – Italy –
    History – 20th century. 3. Science and state – Argentina – History – 20th century.
    4. Dictatorship – Spain – History – 20th century. 5. Dictatorship – Italy – History –
    20th century. 6. Dictatorship – Argentina – History – 20th century. 7. Spain – Politics
    and government – 1939-1975. 8. Italy – Politics and government – 1922-1945. 9.
    Argentina – Politics and government – 1955-1983. I. Gómez Rodríguez, Amparo.
    II. Canales Serrano, Antonio Francisco, 1966- III. Balmer, Brian, 1965-
    Q127.S7S38 2015
    338.9'26 – dc23   2014049925

ISBN 978-1-4724-2232-3 (hbk)
ISBN 978-0-367-59830-3 (pbk)

Contents

List of Figure and Tables   *vii*
Notes on Contributors   *ix*
Preface   *xiii*
Amparo Gómez, Antonio Fco. Canales and Brian Balmer

1   Science Policy under Democracy and Dictatorship: An
     Introductory Essay   1
     Amparo Gómez, Brian Balmer and Antonio Fco. Canales

2   The 'Social Contract' for Spanish Science before the
     Civil War   27
     Amparo Gómez

3   Spanish Science: from the Convergence with Europe to
     Purge and Exile   59
     Francisco A. González Redondo

4   The Reactionary Utopia: the CSIC and Spanish
     Imperial Science   79
     Antonio Fco. Canales

5   Broken Science, Scientists under Suspicion. Neuroscience
     in Spain during the Early Years of the Franco Dictatorship   103
     Rafael Huertas

6   Cultures of Research and the International Relations of
     Physics Through Francoism: Spain at CERN   121
     Xavier Roqué

7   The National Council for Research in the Context of
     Fascist Autarky   141
     Roberto Maiocchi

8   Statistical Theory, Scientific Rivalry and War Politics in
        Fascist Italy (1939–1943)   159
        Jean-Guy Prévost

9   Science, Military Dictatorships and Constitutional
        Governments in Argentina   179
        Pablo Miguel Jacovkis

10   Science Policy in Argentina During the 'Dirty War'   199
        Diana Maffía

Appendix: History of Science in Spain, Italy and Argentina   211

Index   *215*

List of Figure and Tables

Tables

6.1. Data concerning expenditure on basic research, applied
      research, development and services (in million pesetas)   127

6.2. Estimate of research expenditure in public establishments
      for some scientific and economic sectors   127

6.3. Spain's contribution to CERN, 1961–71, in million pesetas   129

Figure

8.1   A three-level structural homology   174

Notes on Contributors

Brian Balmer is Professor of Science Policy Studies in the Department of Science and Technology Studies, University College London. His research interests are in the history of science policy, scientific migration and the 'brain drain', the geography of knowledge, and the relationship between science and the military. He has published widely on these subjects and is author of *Britain and Biological Warfare: Expert Advice and Science Policy 1930–65* and *Secrecy and Science: A Historical Sociology of Biological and Chemical Warfare*. He is an editorial advisor for *Social Studies of Science*, and until recently was on the editorial board of *Notes and Records of the Royal Society*.

Antonio Fco. Canales is Lecturer of History of Education at the Universidad de La Laguna. He is a specialist in the history of the Franco regime. He is the author of books as *Las otras derechas* (Madrid, 2006) and the editor with A. Gómez of *La larga noche de la educación española. El sistema educativo español en la posguerra* (Madrid, 2015); *Estudios Políticos de la Ciencia* (Madrid,, 2013) and *Ciencia y fascismo: la ciencia española de postguerra* (Barcelona, 2009). He has published articles in journals such as *Gender and Education, History of Education* and *Journal of War and Cultural Studies*. He is a member of the Executive Committee of the Spanish History of Education Society and the secretary of its journal, *Historia y Memoria de la Educación*.

Amparo Gómez is Professor of Philosophy of Science in the Universidad de La Laguna and Honorary Senior Research Fellow in the Department of Science and Technology Studies, University College London. Her research interests are in the history and philosophy of science and science policy studies. She is the editor, with E. Agazzi and J. Echeverría, of *Epistemology and the Social* (Amsterdam, 2008), and, with A.F. Canales, of *La larga noche de la educación española. El sistema educativo español en la posguerra* (Madrid, 2014); *Estudios Políticos de la Ciencia* (Madrid, 2011) and *Ciencia y fascismo: la ciencia española de postguerra* (Barcelona, 2009). She is author of 'The rebels and the new Spanish scientific culture' (*Journal of War and Culture Studies*, 2(3), 321–33, 2009) and many other articles in national and international journals such as *Epistemologia, Teorema, Isegoría, Anthropos-Iatria* and *Dynamis*, among others.

Francisco González Redondo is Doctor in History of Mathematics and Doctor in Philosophy of Science, and Senior Lecturer in the Department of Algebra at the Complutense University of Madrid. He has published with Kendrick Press (USA) and in journals such as the *International Journal for the History of Engineering & Technology*, *Journal of New Energy* and *The Journal of the Airship Heritage Trust*. He is Academician of the Academy of Science and Engineering of Lanzarote and member of the Board of the Spanish Society for the History of Science and Techniques, and has been Secretary of the Society Puig Adam of Teachers of Mathematics.

Rafael Huertas is Research Professor at the Higher Council for Scientific Research (CSIC). In recent years, he has been working on the history of psychiatry in Spain and has published in journals such as *History of Psychiatry*, *International Journal of Mental Health* and *L'Evolution Psychiatrique*. Another research line has focused on health studies during and after the Spanish Civil War and he has published articles in journals such as the *American Journal of Public Health* and the *Journal of the History of the Neurosciences*. He has been head of the Department of History of Science (History Institute – CSIC), and President of the Spanish Society for the History of Medicine.

Pablo M. Jacovkis is a mathematician specialised in interdisciplinary computational modelling. For many years he has been a consultant in mathematical models in fluvial hydraulics, hydrodynamics, meteorology and geology, subjects on which he has also published many peer-reviewed papers in international journals. At the University of Buenos Aires (UBA) he was Director of the Department of Mathematics at the School of Engineering and Director of the Institute of Applied Mathematics, Academic Secretary (Provost) and Dean of the School of Sciences. He was President of the National Council for Scientific and Technological Research. Currently he is Vice-President for Research and Development of the National University of Tres de Febrero and Professor Emeritus at UBA. In recent years he has also worked on the history of computer science and applied mathematics in Argentina, and on the links between democratic and authoritarian governments, science and technology.

Diana Maffía is Doctor in Philosophy (University of Buenos Aires), Senior Lecturer at the University of Buenos Aires, and researcher at the Interdisciplinary Institute of Gender Studies of the UBA (she has also been Lecturer at the National University of Rio IV, and University of La Plata). She is Academic Director of the Hannah Arendt Institute of Politics and Culture and she is author of many books and articles. She has been Human Rights Ombudsman (1998 to 2003) and MP at the parliament of the City of Buenos Aires (2007 to 2011).

Roberto Maiocchi graduated in electrical engineering and is Doctor in Philosophy of Science. He is Professor of History of Science at the Catholic University of Milan. He is the author of the books *Scienza e fascismo*; *Storia della scienza in Occidente. Dalle origini alla bomba atomica*; *Elementi di fisica del Novecento*; *Libro: Scienza italiana e razzismo fascista*; *Gli scienziati del Duce*; *Il determinismo nella scienza moderna*; *Ascesa e declino della scienza moderna* and *Anche il caos ha le sue regole*.

Jean-Guy Prévost is Professor of Political Science at the University of Quebec at Montreal. He teaches and conducts research in the fields of the history of political ideas, political theory, the socio-political sciences and research methodology. He is particularly interested in the history and sociology of statistics, a subject on which he has published a number of books and articles, notably *A Total Science. Statistics in Liberal and Fascist Italy* (2009) and, co-authored with Jean-Pierre Beaud, *Statistics, Public Debate and the State, 1800–1945* (2012).

Xavier Roqué is Senior Lecturer in History of Science at the Universitat Autònoma de Barcelona (UAB), where he also directed its Center for the History of Science between 2000 and 2010. His research deals with the history of the modern physical sciences, the material culture of science, and the relations between science and industry. Together with Soraya Boudia, he edited *Science, medicine, and industry: The Curie and Joliot-Curie laboratories*, a special issue of *History and Technology*, 13, no. 4 (1997). His publications also include 'The Manufacture of the Positron', *Studies in History and Philosophy of Modern Physics*, 28 (1997): 73–129. He is currently working on the relation between physics, culture and politics in contemporary Spain, and has recently edited, together with Néstor Herran, *La física en la dictadura. Físicos, cultura y poder en España, 1939–1975* (Bellaterra, 2012).

Preface

Amparo Gómez, Antonio Fco. Canales and Brian Balmer

The aim of this book is to investigate key aspects of scientific and technical development in three countries in Southern Europe and America: Spain, Italy and Argentina. The book focuses on the policies, institutions, science and technology developed in these countries from the late nineteenth century to around the 1970s. It is, therefore, a book about the historical relationship between science and politics in three countries that had significant similarities between their efforts to develop modern science and technology. The book shows how these countries strove for scientific and technical development similar to the most scientifically advanced countries, although this effort took place in a very complex scenario of political upheaval, characterised by revolutionary attempts, periods of dictatorship and brief democratic phases. Although not unique to these countries, this context determined, in ways that the authors in this book try to unravel, their scientific development. This makes it possible to establish interesting parallels among these nations.

This collection is not primarily an account of major scientific or technological discoveries, but instead studies interrelationships between science and politics throughout the twentieth century in these countries. As such, the contributions that make up this volume share the assumption that social, economic and political factors influence scientific and technical development. Some of the authors would push this claim further, contending that a division between 'external' and 'internal' factors is misleading because the two are co-constructed. Regardless of these slight differences, because the chapters all draw wider factors into their analysis of the three countries studied, they are likely to capture elements of what has happened in many other Southern countries. Furthermore, despite telling histories of national science, Spain, Italy and Argentina are all part of a wide geographical region and so it is possible to draw comparisons between the three countries. It is also important to clarify that the focus of each chapter, each of which is on national science, does not ignore the fact that science and technology, although produced in specific places, have important international and transnational dimensions.

That said, our book does not intend to be a systematic comparison of transnational issues in the three countries. Certainly, transnational considerations appear, particularly the movement of peoples and scientific ideas across national boundaries. But the authors address questions from different scientific traditions and focus on different states of research in their individual chapters. An attempt to unite them into an overarching transnational perspective would be, at best, presumptuous. More modestly, but realistically, the book studies the relationships between science and politics in these three countries, stemming from the belief that they are significant cases of undoubted interest for the history of science.

The relationships between science and politics have been central to modern scientific and technical development. Governments of countries such as the United Kingdom, France, Germany and the United States, but also Italy, Spain and Argentina, have long supported scientific and technological research and promoted scientific institutions. In the twentieth century, science and technology played a key role in the two world wars, and governments in many countries developed science policies that gave impetus to science and technology with greater or lesser success. Subsequently states continued to intervene in scientific and technical development and invested significant financial resources with the goal of increased economic and military competitiveness for their countries.

It is worth noting here that we have used 'science policy' in this book in a deliberately broad and loose way, as a heuristic to denote any deliberate government intervention into the creation, organisation, development or even destruction of science and technology. It also denotes the involvement of science and scientists in government. While this definition might open us to accusations of either anachronism or, conversely, of ignoring contemporary emphases on governance (rather than government), and the role of non-state actors, in the absence of an analysts' definition of 'science policy' that is universally agreed on, we remain comfortable with our terminology.

State intervention in science was crucial for the scientific and technical development of Europe and America, although such development did not happen either on the same terms or with the same results in all countries. This meant, among other things, the existence of a major North–South division in science (and in Europe also regarding Eastern countries). The reasons why state intervention in the scientific and technical development of Southern European and American countries had comparatively limited results and at certain times failed partially, or completely, are fundamental to explain the type of development achieved by these countries in science and technology and understand the existence of such a North–South borderline.[1] The contributions of this book discuss and try to explain these reasons through the analysis of what happened in the case of Spain, Italy and Argentina. While acknowledging this boundary, we have deliberately avoided the notion of 'peripheral science', with its connotation of a one-way flow of information and resources from a 'core' of 'good science' to a 'periphery' of 'at best shoddy science', throughout the book.

The political history of science has been very fertile and produced a significant body of work on science and technology in the United States, Germany, the United Kingdom, France and the former Soviet Union (and, in recent years, Eastern Europe could perhaps be added to this list). This productivity also reveals that the political history of science has largely been written by authors belonging to the academic and research communities of Northern Europe or America, and so is fundamentally a history of science and technology in countries with relatively strong scientific-technical development. Southern countries have hardly been considered, even by the most recent histories of science that do pay attention to other geographic areas such as China, Japan and India. Such absence seems to indicate that Southern countries have provided nothing of interest for historians of science and technology, and that what has been done in these countries to develop science and technology is of little significance. From this perspective the Southern countries only contribute sporadically to the History of Science with some occasional figures awarded Nobel prizes and some very limited discoveries. However, this book demonstrates that these claims are difficult to sustain. The countries of Southern Europe and America developed interesting and often successful policies and have produced valuable science and technology. The 'sin' of these countries is that their scientific and technical production has been much lower than in the North and, although good science policies were proposed over the twentieth century, economic investment in science and technology has been well below that of rich Northern countries.

Therefore, we hope this collective work will fill an underdeveloped space in the History of Science. It makes no claim to be comprehensive. It does, however, respond to a wider interest to understand the history of science and

---

[1]   This terminology is widely used and very useful for this book. We could talk about 'more developed and less developed countries', 'central and peripheral countries', or 'Latin countries' etc. But, while the countries that are labelled Northern or Southern might change over time, for the period we are considering the North–South terminology is very clear geographically and also politically and economically, and includes the meanings of other terms as poor–rich, developed–less developed, etc. (See for example, *Science, Technology and Human Values*, Nov 2014 39(6) Special Issue: Voices from within and Outside the South – Defying STS Epistemologies, Boundaries, and Theories).

technology in countries whose scientific and technological development was not canonical, in order to provide a plural and polycentric picture of Western scientific and technical development.[2] At a more theoretical level, the book indirectly addresses important concerns about sites and locations of science that have recently emerged in the historical geography of knowledge.[3] In the rest of the book authors show how dominant scientific ideas penetrated into Spain and Italy; how the topic of the importance of technology to the development of countries was present in the three countries; how scientists were moving from their countries to the great international research centres for training; how in the fascist context the idea of autarky was moving between countries and was accepted; and how ideas about institutionalising science and the emergence of large scientific institutions, were shared by the three countries.

The book also contributes in other ways at a more theoretical level, with the chapters touching on key topics such as the role of science policy in countries that have experienced both dictatorial and democratic regimes; new forms of organisation and institutionalisation of science and the role of scientific institutions in the implementation of policies in democracy and dictatorship; the involvement of scientific communities in the governance of science and its institutions; the level of political involvement of scientists in democratic and non-democratic contexts; the role of ideology in scientific organisation and research; the scientific practices adopted by scientific communities in different contexts; and the characteristics of science and technology produced in these contexts.

And what, finally, of the global context? Although we do not pretend to take on the enormous task of explaining dictatorships and science policy in a general international context, within their analyses the contributors to this collection take into account various dimensions of the international context relevant to their own analysis. These include: the development of Italian imperialism during fascism; the importance of the Second World War; the influence of the new international context after the Second World War; international exile; the Cold War; international science; international nuclear energy policies; and the questioning of Northern science from a Southern anti-colonialist perspective in Argentina.

---

[2]   For a review see Cozzens, S., 2008. Science and Development. In: E.J. Hackett, O. Amsterdamska, M.E. Lynch and J. Wajcman (eds), 2008. *The Handbook of Science and Technology Studies*. Cambridge: The MIT Press.

[3]   For recent overviews see: Powell, R., 2007. 'Geographies of science: histories, localities, practices, futures'. *Progress in Human Geography*, 31(3), pp. 309–29; Naylor, S., 2005. 'Introduction: historical geographies of science–places, contexts, cartographies'. *The British Journal for the History of Science*, 38(1), pp. 1–12.

In his book, *Science in the Twentieth Century and Beyond,* Jon Agar uses the metaphor of 'isolated spots of light in a dark night' to describe the current state of the history of twentieth century science. This is an excellent metaphor of the history of science in the countries of South America and Europe, in which there would be no History of Science beyond some occasional flashes. But of course, this large geographical region also has a history of science and technology, and the History of Science (with capital letters) of the twentieth century cannot be understood without it. In this way, we aim to modify, albeit slightly, this asymmetry in the history and geography of scientific knowledge. This objective is particularly relevant at a time when the gap in scientific and technological development among European Northern and Southern countries increases alarmingly amid the severe economic crisis the South suffers. This growing inequality is having the consequences, among others, of reducing the total production of research in Europe and generating a 'brain drain' from Southern to Northern Europe.

The Contents of this Book

An introductory essay by Amparo Gómez, Brian Balmer and Antonio Fco. Canales acts as a framework, helping to understand the cases beyond their particularities. In this chapter some of the specific issues set out in the chapters are related to a broader theoretical and historiographical context. The rest of the book is divided into three parts corresponding to each of the aforementioned countries.

The first part of the book focuses on Spanish science and consists of five chapters. The section begins with the chapter by Amparo Gómez titled 'The "Social Contract" for Spanish Science before the Civil War'. In this chapter, the scientific, intellectual and political movements that made possible a 'social contract' for science in Spain between 1907 and 1939 are analysed. This contribution describes Spanish science policy and the institution that embodied this 'social contract' for science, and analyses its scientific and technical achievements. It also demonstrates that this contract was subject to several boundary and integrity problems that were made explicit in political debate. Finally, it indicates how the Civil War and Franco regime broke the social contract for science. Francisco A. González studies this radical breakup in his chapter 'Spanish Science from the Convergence with Europe to Purge and Exile'. The author shows how the victory of Franco in the Spanish Civil War finished the processes of scientific convergence with Europe and analyses the impact of victory on the scientific communities through purge and exile. In 'The Reactionary Utopia: the CSIC and Spanish Imperial Science', Antonio Fco. Canales addresses policy towards science developed by the Franco regime after the victory in the Spanish Civil War. The author provides a characterisation of the Higher Council for Scientific Research as the result of the adoption of a reactionary utopian vision: overcoming the dichotomy between faith and reason through the restoration of a Spanish Imperial Science. Rafael Huertas, in 'Broken Science, Scientists under Suspicion. Neuroscience in Spain during the Early Years of the Franco Dictatorship', explains the scientific and institutional vicissitudes of neurosciences after the Civil War. The aim of his chapter is to explain why Francoism proceeded to the 'political cleansing' of the internationally prestigious *Instituto Cajal*, and how this impacted on health research. The chapter sets these events in the context of the hunger and penury of the post-war years that resulted in a high incidence of deficiency neuropathies and pellagra, and an epidemic of neurolathyrism. Finally in this section, in the chapter 'Cultures of Research and the International Relations of Physics Through Francoism: Spain at CERN', Xavier Roqué explores the international relationships of Spanish physicists during the Franco regime and especially Spain's first stage at the European Organisation for Nuclear Research (1961–68). The author argues that, beyond the lack of political will and the economic burden that are usually blamed for Spain's withdrawal, there were 'failures of communication' between CERN and Spain. These failures were an expression of the scientific culture set by the winners in the Civil War in the 1940s that had survived into the 1960s.

The second part of the book focuses on Italian science and it starts with Roberto Maiocchi's chapter 'The National Council for Research in the Context of Fascist Autarky'. In this work Maiocchi studies the history of the National Council for Scientific Research, the most important Italian scientific institute in the years of Fascism, until the Second World War. The focus is on the role of the institute in carrying out the autarky project, which involved the whole of Italian society from 1935 onwards. It is argued that the National Council for Research would eventually prove to be unable to reach the goals set by the political powers. Jean-Guy Prévost, in the chapter entitled 'Statistical Theory, Scientific Rivalry and War Politics in Fascist Italy (1939–1943)', takes into account the institutional – by contrast with the narrowly scientific or conceptual – development of Italian statistics, and the personal, domestic and transnational rivalries among statisticians. Prévost analyses Gini's attempt to provide an alternative statistical orthodoxy for a post-war new world order in which Anglo-Saxon hegemony over statistics and science in general would have perished as a correlate of the Allies' defeat on the battlefield.

The third part of the book focuses on Argentinian science and starts with Pablo Miguel Jacovkis's chapter 'Science, Military Dictatorships and Constitutional Government in Argentina'. In this chapter the author examines the evolution of science policy and scientific-technical production in Argentina in the turbulent context of alternating dictatorship, constitutional and populist periods. In this study attention is paid to civil policy, but also to the role of the Argentinian Army in relation to both the successes and failures of developing modern science and technology in this country. Diana Maffía, in 'Science Policy in Argentina during the "Dirty War"' analyses, through her personal experience, the evolution of scientific training and research practice conducted within the Argentinian universities during the different dictatorships that this country endured from 1930 onwards. Simultaneously, this chapter seeks to show that the science policy of Argentina's last military dictatorship had its antecedents in the science policy developed in previous constitutional periods.

Acknowledgements

We would like to thank to Emily Yates and Kirsten Weissenberg, our editors at Ashgate for their support for this book. Also, thanks to the series editors Robert M. Brain and Ernst Hamm for their support and encouragement. We thank the anonymous reviewer who provided us with extensive and thorough feedback, for devoting much time and effort to the first draft of this book. Diana Maffía would like to acknowledge Dr. Beatriz Kohen and MSc Aluminé Moreno for the translation of her chapter. We could not have completed this book without the support from the Spanish Ministries of Science and Innovation, and Economy and Competitiveness (National Research Projects FFI2009–09483 and FFI2012–33998) and from the Universidad de La Laguna. Thanks also to the Department of Science and Technology Studies at University College London for supporting this project, particularly through a visiting research fellowship for Amparo Gómez and a period of sabbatical leave for Brian Balmer.

Chapter 1

Science Policy under Democracy and Dictatorship: An Introductory Essay

Amparo Gómez, Brian Balmer and Antonio Fco. Canales

The complex and shifting relationships between science and politics will be considered from quite different historical perspectives in the course of this book. The aim of this chapter is to provide a broad framework that introduces the main themes explored in detail in the individual chapters. Whether each chapter takes a broad historical sweep or focuses on specific disciplines, this book reinforces the argument – commonplace within contemporary historical and social studies of science – that politics and science deeply affect one another. Moreover, despite often being labelled as less scientifically developed, the three countries considered in this collection all seriously regarded science and technology as key to their more general development.

This chapter opens with a discussion of how we conceptualise the relationship between science and politics. In particular, it tackles the difficult question of whether democracy is a pre-condition for good science. This leads us to a discussion of the characteristics of science policy in specific political regimes: first science policy and democracy, and second science policy and dictatorship. The final section turns from policy intervention into science and instead focuses on the various different ways scientists, as individuals or as a collective, adapted and changed under shifting political conditions.

Science and Politics

A significant issue in this collection is whether it is possible to identify different types of relationship between science and politics – and then between politics and scientific development – depending on countries and political regimes. This is not a trivial topic because it is often assumed that science and technology develop better in democracies than in dictatorships, since there is a relation between scientific research and democratic values (Kitcher, 1993 and 2011). Thus, certain values such as 'freedom', 'honesty' or 'criticism' have often been considered characteristic of modern science and, at the same time, of liberal-democratic regimes. These conditions are expressed by Merton's *scientific ethos* (1973) or Polanyi's *dynamic orthodoxy* (1969) and the values defended by classical philosophy of science, for example, by Karl Popper (1945). In democratic contexts science would be independent of policy, and rational, honest and critical scientists would produce and govern science. Therefore, it is argued, science flourishes in democracy with great intellectual freedom and a minimum of political and ideological control (Goudsmit, 1947).

Freedom is essential since it would be inherent to science and without it neither objectivity nor scientific truth would be possible – both would require the absence of external control and thus freedom is a condition of possibility. Scientific integrity would be another characteristic of scientific practice which would largely be dependent on the autonomy of a scientific community distanced from political management. As the chemist and philosopher Michael Polanyi (1962) affirmed, scientific development needs an autonomous republic of science, free from external political intervention.

This model of relations between science and policy is captured by the notion of the 'Social Contract for Science', defended by scientists and incorporated into the science policies of different countries in the twentieth century (Bush, 1945, Guston and Keniston, 1994). The social contract is not a written contract. It assumes a tacit understanding that the state supports science and leaves it to act independently in producing new socially beneficial products. As will be discussed in Gómez's chapter, Spain in the early twentieth century developed a science policy that, in essence, was a social contract for science. The same happened in other countries, as was the case for British science policy since the early twentieth century with the assumption of the Haldane principle.[1]

Research under social contract, at least theoretically, makes it possible to detect and ultimately prevent scientific fraud and the development of pseudoscience, and thus, reject bad science. Indeed, in this respect, Merton's norms could be

---

[1]   The so-called Haldane principle, as a social contract for science, involves independence between science and policy and that decisions about research and funding distribution are taken by researchers rather than politicians. It was derived from The Haldane Report (1918) *Report of the Machinery of Government Committee under the chairmanship of Viscount Haldane of Cloan*. London: HMSO. Edgerton (2009) argues that the term 'Haldane principle' was only coined in the 1960s, but adds that prior to this: 'it was the case however, that in the realm of education, and non-departmental research, there was a powerful sense of the limits of appropriate government intervention'.

construed as an ethical underwriting for the epistemic dimension of science. The wider idea is that science and democracy reinforce each other, and that good science is possible in democracy where scientific standards are fulfilled. On the contrary, under dictatorships science and technology would be subject to the objectives of the regimes and would be permeated by ideology. In this type of political regime science would be distorted and highly politicised; scientists lose their autonomy and become involved in politically driven projects.[2]

Nevertheless, while the picture painted above is both reassuring and comfortable, the history and sociology of science has shown that this dichotomist conception – democracy good science, dictatorship bad science – is over-simplistic. Historically, deviations from the social contract for science and the values that it implies were identified in democracies in at least two areas: a) attempts by politicians for greater intervention in science than those suggested by the contract, and b) problems about either the integrity or productivity of scientific communities that challenged their claims to autonomy (Guston and Keniston, 1994 and Guston, 2000).

Boundary problems between science and policy were generated by governments attempting to intervene in science. It is not difficult to find historical examples that show greater policy intervention than suggested by the social contract for science.[3] During the Cold War, much science was strongly influenced by the defence priorities of governments and the needs of the military, with much work carried out in secret (see for example Forman, 1987; Edgerton, 2006; Balmer, 2012; Wolfe, 2012). In the United Kingdom the Haldane principle was revised in the early 1970s by the Rothschild Report. The Report embodied the so-called 'customer-contractor' principle whereby a proportion of funds that had previously been allocated to the quasi-governmental, and therefore supposedly autonomous, research councils for distribution to the scientific community, were now allocated directly to government ministries to request specific, targeted research. Regardless of the actual impact of Rothschild, the customer-contractor principle was widely held as a significant shift away from the autonomy of science and towards more interventionist, directed modes of support for science (Gummett, 1980). The debate about the effects of the social contract for science and the Haldane principle remains open (Calver, 2012).

---

[2] For example, as Szöllosösi-Janze (2001, p. 4) points out in the German case, it was thought that National Socialism was not a very scientific regime (at least before the Second World War) and the Third Reich promoted pseudoscience and kept qualified scientists separated making it impossible for them to carry out their work.

[3] For more on this topic see Gómez (2014).

If political intervention in science was an early temptation of a state which was never fully benevolent, the second type of problems that the social contract for science faced was generated by scientific communities. Scientific practice did not always conform to the contract nor to the standards on which it was supposedly based. Problems with integrity and productivity soon appeared, but became more abundant over time. These had a considerable impact on the political and social perception of the scientific community. So, for instance as early as 1939, J.D. Bernal argued that many things, like social goods, were more important than researchers' freedom (Bernal, 1939).

To deviations of the social contract in democracy we must add another important though uncomfortable fact, the recognition that successful science and technology are possible under dictatorial regimes, as happened in the case of the countries studied in this book and others, such as in Nazi Germany or the Soviet Union (Szöllosösi-Janze, 2001; Saraiva and Wise, 2010; Walker, 1995, 2001; Proctor, 1988; Graham, 2004). As pointed out by Saraiva and Wise (2010, p. 421): 'Scholarship on Nazi and Soviet science no longer deals exclusively with scientific exiles or perversions of the scientific method. Mengele and Lysenko now share the stage with a multitude of normal scientists'. Under these regimes, sound science and technology were produced (of course, not invariably), although research was pursued in ways and with objectives marked by the policies of the regimes, and political and ideological interference affected the conditions of scientific–technical production more directly, and more violently, than in democracies.

There have even been paradoxical situations, such as in Argentina between 1955 and 1966. As Jacovkis describes in his chapter, under a military government from 1955 to 1958 scientists and professors who had been repressed by the previous Peronist government returned to their jobs, the university was democratised and it began to reach a level of development that was considered the 'golden age' of universities. It was in this period that Argentinian basic science began to flourish and important national scientific institutions were founded. But this took place only under this government, not in other dictatorial periods, neither in most of the constitutional stages, including much later in Menem's government. Also, in Mussolini's Italy, modern scientific and technical development was supported by the State looking for technological and scientific effectiveness.

On the other hand, there is no doubt that under dictatorships there were cases of pseudoscience, distortion of scientific practice and even fraud. In Francoist Spain, Antonio Vallejo Nágera's research with war prisoners to establish the biological foundation of their leftist ideologies was one well known example (Vallejo, 1938, 1939a, 1939b and 1939c; Vallejo and Martínez, 1939;

Nadal, 1987; Vinyes, 2001; Richards, 2012).[4] Prominent fraud occurred in 1940 with the project aimed at creating petroleum by mixing water with herbs and other secret products, and later with an attempt to produce synthetic gasoline with coal and bituminous slate (Preston, 2008).[5] Nevertheless, in non-dictatorial regimes, although the conditions in which science is developed assume degrees of scientific independence that were not present under dictatorships, different problems and fraudulent activity also occurred. Under Perón's elected government, there was the cold fusion fraud, the most famous and most expensive case of scientific fraud in Argentina.[6] Then in Spain, as Gómez notes in her chapter, and although not an attempt to deceive people about scientific findings, we find that networks of influence introduced biases to the heart of the Board for Advanced Studies and Scientific Research (JAE). Science and technology produced both in democracy and dictatorship have broken some of the principles that supposedly govern the production of good science. Therefore, the conditions of the Mertonian *ethos* or the requirements of Polanyi for scientific activity are not indispensable for successful scientific and technical development. Neither, contrary to Merton's theory, is it the case that when norms are violated that a professional self-regulatory system of sanctions occurs. Instead, these normative conditions constitute an idealisation that offers, at best, a set of aspirations and, at worst, a professional ideology that presents a purist and unrealistic image of science (Mulkay, 1979). So, to summarise, it is tempting but mistaken to claim that science is only possible in a liberal or democratic context. Science and technology have been developed with success in totalitarian contexts, and even by not very democratic scientific communities. This is by no means a defence of science under dictatorships, but an attempt to write histories of science that do not start from the pre-supposition that science can only work in democracies (Saraiva and Wise, 2010).

Nevertheless, not all influences are the same, and the specificity of how dictatorships affect science and technology might be found mainly in the political–academic conditions of research, affecting the practice of science, and

---

[4]   Antonio Vallejo Nágera, as commander of Military Psychiatric Services and director of the Office of Psychological Research Concentration Camps, had a privileged position to investigate this issue. Vallejo had access to a significant number of Civil War prisoners who he divided into groups that brought together all the elements considered to be subversive or problematic, from Marxist–Communist guerrillas to Basque and Catalan separatists.

[5]   *La Vanguardia Española*, 8 February 1940.

[6]   The Austrian scientist Ronald Richter promised Perón that he was going to get nuclear fusion under controlled laboratory conditions. Perón believed him and spent a great deal of money on this project (we thank Professor Pablo Jacovkis for this information).

in the political and ideological atmosphere in which scientific communities develop their activity. Consequently, instead of the polarity of democracy equals good science and dictatorship equals bad science, it should be taken into account that between periods of democracy and dictatorship – in the case of science and technology – there are both ruptures and continuities. The clearest examples of continuities in this collection are found in Maffía's chapter, which shows how legislation and attitudes developed in constitutional periods in Argentina were so antidemocratic that they continued to be applied usefully under the most terrible dictatorship. The clearest example of rupture, as discussed by various contributors here, is Spanish science and technology before and after the Civil War.

A more nuanced way of thinking about how both democracies and dictatorships might have specific effects on science is to use Harwood's notion of national styles of science. Harwood (1993) argues in his study of the development of scientific ideas that there are national styles of science defined by different values, norms, assumptions, research traditions and funding patterns. From this perspective Prévost's account of Italian statistics in this collection could be construed as the development of a particular style of research that conformed to the requirements of Italian Fascist goals. Likewise, Roqué's chapter clearly demonstrates how the requirements of Franco's Regime produced physicists working in a research culture that was significantly different from that of physicists working at CERN.

Science and Democracy

Some chapters in this book deal with science and technology under parliamentary and constitutional governments in Spain, Italy and Argentina. In the nineteenth century, under such governments, the scientific and technical development of these countries were each quite distinct: rather limited in Spain, most apparent in Argentina and it went through different stages in Italy before and after unification. What these countries had in common, however, was that their governments began taking an interest in science and technology. Significantly, what did exist was a political perception that science and technology might be important for the development of each country (although this hope was not shared by the entire political class). In Spain, where this view was held by politically progressive governments, no steps were taken to give political support to the development of science and technology because there were other political priorities. In Argentina the idea of science and technology at the service of nation-state building was fundamental. Italy, before unification, had cosmopolitan scientific activity in some areas, which facilitated good relationships with other European countries (Benzi, 2010).[7] As pointed out by Pancaldi (1993, p. 23): 'The Royal Society, the Linnean Society, and the Astronomical and Geological Societies of London elected Italian scientists as foreign members'. Such cosmopolitanism coexisted with nationalism and many scientists played a role in the birth of the new Italian State, proposing measures to defend a centralised policy following the French model. Since the end of the nineteenth century Italian science policy aimed to unify its universities and scientific institutions as the resources for teaching and scientific research decreased.

Italy and Spain left the nineteenth century developing national policies to improve scientific and technical development. Different measures were formulated that aimed to elevate the level of scientific and technological development in each country. In the case of Italy this objective focused primarily on technology (without neglecting science), while Spain first targeted basic science (and technology on a smaller scale). Therefore, given that its industrial development had started, at the beginning of the twentieth century Italy was especially interested, and invested resources, in developing applied research. Italian science and technology was developed in the First and Second World Wars and the military effort increased the science budget. Scientific institutions cooperated internationally and Italian scientists had come into contact with the recently established national laboratories. For example, 'Volterra was impressed by the National Physical Laboratory at Teddington' (Pancaldi, 1993, p. 35).

Spain developed its science policy in the early twentieth century in basic science and later in technology. The relationships between Spanish governments and science and technology resulted in a social contract for science. Although the contract had to face different problems, as Gómez's chapter shows, the ideals of freedom, independence and productivity animated Spanish science policy until the Civil War in 1936. This policy helped the country achieve a level of scientific and technical development not reached before or after, for decades, in Spain. By the outbreak of the Civil War, Spanish science had reached a significant level of development, and had begun to close the gap with European science. The intrusion of Fascism resulted in a bloody Civil War which also represented a radical rupture in the country's scientific development; what had been achieved in three decades collapsed with the War and the subsequent Franco Regime. By

---

[7]   Before Unification, Italy had important physicists and chemists such as Luigi Galvani, Alessandro Volta, Amedeo Avogadro and Stanislao Cannizzaro. They were also great mathematicians.

comparison, in the case of Italian Fascism from 1923, it did not entail this type of break in scientific development. This was an essential difference between the two countries.

Argentina, like Italy, began the twentieth century interested in developing technology rather than basic science, since it was understood that technology was essential for industry. Spanish industry did not need much technology and innovation, and when it was required, it was acquired from outside the country because it was more profitable than investing in inventions themselves. However, in Argentina interest in science and technology gradually dwindled. The traditional ruling class at this time was losing interest in both science and technology, illustrated by the fact that they even 'brought the plans of their palaces from Europe' (Jacovkis, 2003, p. 3). They did not understand the institutional problems of science; neither did they pay much attention to science. This neglect was due in part to the fact that 'the extraordinary agricultural wealth of our country and its rapid growth gave a sense of confidence and self-reliance' (Jacovkis, 2003, p. 3), together with a supposition that it was less costly to acquire the scientific and technological 'products' of those countries that had developed them. In these circumstances no great indigenous scientific development was deemed necessary.

The value of the autonomy of science and technology from politics, along with other values associated with science, were questioned in Argentina during its parliamentary and constitutional stages. Scientists were opposed to this situation and defended a policy based on research freedom and autonomy of science from the State (essentially, a social contract for science). But, the tendency of Peronism was to intervene politically through measures, legislation and attitudes that questioned the independence of scientific research and education and that later were applied effectively in the last Argentinian dictatorship, as Maffía's chapter shows. For example, under the elected government of Perón in 1946, persecution and dismissal of scientists was carried out and many scientific careers were interrupted. Scientists could not get work if they did not join the Peronist party (probably the most famous example of this exclusion was that of Mario Bunge, originally a physicist and then philosopher). University autonomy was eliminated and the universities lost a third of their teaching staff.

It is possible also to establish parallels between the Southern countries analysed here and Northern countries such as England, France and Germany. The political intervention of Northern governments in science and technology in the nineteenth century was more substantial than in the Southern countries analysed in this book. Germany and France's public policy involved national planning of research and quite fluid relations with industry (Prost, 1988; Picard and Pradoura, 1988; Fox, 2010; Fjæstad, 2010). In England, the development of industrial support was important but not always well regarded by scientists and industrialists. However there were forms of collaboration such as in the Great International Exhibitions of 1851 and 1862 (Cannon, 1978; Clarke, 2010; Coopey, 2002; Edgerton, 2005; Withers, Higgitt and Finnegan, 2008).

From the beginning of the twentieth century in Europe there was a growing awareness that scientific and technological development should be supported by governments, and this occurred with differences and peculiarities in the United Kingdom, France and Germany, but also in Spain, Italy and (beyond Europe) Argentina among other countries. The need for state intervention in science was reinforced after each World War. War was fundamental for the development of science, since it promoted government intervention and financial support for science and technology (Agar, 2012). Spain's neutrality in both wars meant that Spanish science did not benefit directly from this incentive, at a time in which military needs drove other governments to develop important interventions in science, with increased financial budgets and effective results in the production and national organisation of science and technology.

In general, governments recognised the importance of science and technology early on for industrial development, the war effort, and also for the development of other sectors such as health and even agriculture. But, broadly, the biggest difference between the Southern and Northern countries mentioned was the level of scientific and technical development that each reached.[8]

Dictatorships and Science

The relationships between science, technology and politics under dictatorships are widely discussed in this book. In the cases studied, a significant difference in the attitude of dictatorships to science can be detected between Argentina and Spain, on the one hand, and Italy on the other. Generally, the Argentinian and Spanish dictatorial regimes mistrusted science and scientists and this distrust superseded more pragmatic positions based on the tangible benefits attributed to science (except the 1955–58 period in Argentina). In fact, suspicion did not arise over the technological products of science, but over the principles and ideas that were meant to govern scientific practice, and which challenged

---

[8]   Simple data, such as Nobel prizes won by scientists between 1900 and 1950 in each country, give a clear picture of this difference: Germany 42, France 41, Britain 33, Italy 3, Spain 2 and Argentina 1.

aspects of the military authoritarian mentality and especially conflicted with the religious fundamentalist reactionary thought that was present in both regimes. Catholicism was an essential component in both dictatorships and the Catholic Church had adopted, in both countries, a hostile position to scientific development. Ultimately, the Church in these countries was suspicious of science because it was a product of modernity, an activity based on principles such as the independence of reason and freedom of thought, principles the Church loathed. An alignment of the Spanish and Argentinian dictatorships with fundamentalist Catholicism brought this reactionary approach into a powerful position.[9]

This reactionary dimension was particularly clear in the Spanish case during the first decade of the Franco Regime. As Canales shows in his chapter, the victors in the Spanish Civil War believed it was possible to turn their domestic military victory over democratic and progressive forces into an international intellectual victory over the evolution of European thought during previous centuries. Thus, they proposed a reactionary utopia aimed at re-establishing a Spanish imperial science subject to God and the Catholic religion instead of the rationalism and the materialism that characterised modern science. For this purpose, they benefited from the fierce repression against any dissenting voices inside Spain and the crisis of democracies outside, which they saw as the definitive crisis of the civilisation born in the Renaissance. The approach was not as radical in Argentina, although their leaders shared the same kind of desires. The Argentinian military, however, never intended to go on the offensive to challenge international modern science; instead they adopted a more defensive attitude.

In opposition, Italian and also German Fascism, as modern projects, seemed not to find special reasons to distrust science. However, that does not mean that Fascism accepted that it needed to respect the autonomy of scientific communities. Science, like all other areas of society, was to be coordinated and placed at the service of the Fascist totalitarian project. The relationship between Italian Fascism and science was pragmatic and instrumental, based on the Fascist aspiration that Italian scientists contribute to the autarkic economic

---

[9]   This account suggests an entirely conflict-bound model of the science–religion relationship. While this remains the case for the countries studied in this collection, current historiography acknowledges the relationship between science and religion as more historically and geographically variable (for an introduction to this debate see Dixon, 2008). Also note that, as Canales explains in his chapter, the Catholic Church did not want to dispose of science but instead to tame and subordinate it to theology. For the complex relationship between science and the Spanish Catholic Church after the Civil War, see also Camprubí (2014, Ch. 3).

development project and, ultimately, to creating the material conditions that would allow the regime to face its imperial war aims.[10] So, as Maiocchi shows in his chapter, the Italian National Council for Research (CNR) was the key Italian scientific institution that should have transformed Italian science into an instrument that would be useful for Italy's development and economic and military power. However, despite the funds it received, the CNR was unable to turn this project into reality.

Beyond these different attitudes to science, the three dictatorships shared with all dictatorships the temptation to use their immense power to purge and discipline scientific communities. The outstanding example of this ideological intervention was Spain where the purging process was intense and was added to exile, as is shown in the chapter by González. As Gómez explains in her chapter, scientific communities in Spain had flourished in the previous three decades, but emerged after the Civil War virtually dismantled by exile and purge. The Franco Regime was not content with the scientific community adopting a submissive and prudent attitude, but it instead intended to redefine radically this community according to the narrow reactionary ideological criteria of Spanish tradition.

Thus, a new scientific community was formed via the criterion of their affection to the regime. Purging was ideological and intense, so there was no continuity in scientific communities or their science. The winners in the Spanish Civil War even favoured one scientific school over another, such that the purge was turned on schools and entire scientific traditions, especially those that had benefited from the science policy made through the Board for Advanced Studies (JAE), the National Foundation for Scientific Research and Trials of Reform (FNICER) and the Institute of Catalan Studies. As Huertas explains in his chapter, Ramón y Cajal's school was dismantled, but, at the same time, his prestigious name was kept as the 'denomination' of the research institute. Also according to Roqué, this will to intervene for political reasons lasted until at least the 1960s and can be detected in the reaction of Spanish scientists and officials when Rafael Armenteros, a scientist belonging to a family of exiles, was hired by CERN, an international scientific institution. To summarise, between

---

10   Autarky is the economic model in which a nation must be able to supply itself and satisfy all its needs with a minimum of trade with foreign countries and rejecting foreign capital. It fits perfectly with Fascist interventionist nationalism, which found within autarky the economic doctrine that best engaged with their ideological assumptions (and in Spain with the mentality of the military in power). However, autarkic approaches were not exclusive to Fascism, neither were all fascisms autarkic (for instance, the economic policy of Italian Fascism was not autarkic until the early 1930s).

pre- and post-war science there was a radical fracture that had a high cost for scientific and technological production and the Spanish scientific community.

There were also several ideological interventions into Argentina's scientific communities under the different dictatorships, although the situation does not seem as systematic as in Spain. Jacovkis explains in his chapter how, before the purge and disappearances of the last Dictatorship, the elected Peronist government conducted radical ideological intervention which led many scientists into exile and how, paradoxically, the return of these scientists and the 'golden age' of Argentinian universities began to take place under a military government. Maffía insists on the continuity between the constitutional years under the government of Perón's widow and the last Dictatorship. In fact, it seems difficult to conceive of quite how scientific activity could develop in Argentina in this ongoing process of destruction and reconstruction.

By contrast, continuity and stability in the scientific communities was remarkable under Italian Fascism. Actually, the Italian purge (as in Germany) was basically racial and only secondarily political or ideological (Morente, 2005; Vittoria, 1991; Israel and Nasasi, 1998; Goetz, 2000; Boatti, 2010). The vast majority of teachers and researchers were not affected by it and, therefore, there was greater continuity in the scientific community and, above all, a chance to accommodate to the new situation. With few exceptions, the future of researchers depended on their current actions and attitudes and not their past. The purge was a limited and late process. University professors were affected by government Decrees in the 1920s that permitted the removal of public servants for political reasons. Although a few teachers were dismissed for political reasons and others went into exile – to a degree not yet quantified – it is certain that even in 1930 the dismissal of an academic for political reasons raised legal difficulties. This infuriated the Fascist party, which considered that the Fascist revolution had hardly affected the universities. In 1931 progress was made in fascistisation with the obligation of an oath to the Fascist regime. However, this measure had no significant effects on the continuity of teachers, as the majority took an oath for pragmatic reasons and at the recommendation of Catholic, socialist and communist parties. Indeed, only 12 out of 1,213 teachers refused and were dismissed. The contrast with Spain is evident. In this country 12 professors were executed, among them three Rectors (a quarter of the total number of Rectors), and a third of the professors were purged (Morente, 2015, pp. 192 and 194–5). In contrast, the continuity for the scientific community in Italy was strong, at least until the racial laws of 1938. The application of these racial laws resulted in the expulsion of one tenth of the number of professors (Israel and Nastasis, pp. 251–8), still a far cry from the situation in Spain. In general terms, intellectual and scientific autonomy in Fascist Italy was widespread. Obviously, it was not possible openly to criticise the regime or the government, but the continuity of the non-Fascist scholars made it difficult for the Fascists to control knowledge. Moreover, even the Fascists themselves did not want to suppress academic freedom; on the contrary, they granted a broad autonomy to culture, science and thought.

In Germany as well, the purge was eminently racial (Beyler, Kojevnikov and Wang, 2005; Macrakis, 1993). In April 1933 campaigns of harassment against scholars were channelled by the regime through the Civil Service Restoration Act, providing for the dismissal of public employees who were not Aryans or who were deemed politically unreliable (Macrakis, 1993, p. 53). The purge focused mainly on the former, while there were few non-Jewish teachers purged. In universities this purge raised very little opposition from scholars, which shows the speed and ease with which the German universities settled into the new regime. Roughly 16 per cent of teachers were dismissed, for the most part Jewish. The intensity of dismissal, however, varied widely between universities, based on their previous degree of openness to Jews, democrats and socialists. In the universities of Berlin and Frankfurt more than 30 per cent of teachers were removed, while in Leipzig, Königsberg and Marburg the purge just reached about 10 per cent (Grütnner, 2005, p. 93).[11] Overall, it is estimated that over three thousand scholars, of whom about 750 were professors, left the university sector. The impact on national science and technology can be evaluated by considering that they included 24 scientists who were already, or would be in the future, Nobel laureates.

With only relatively slight ideological intervention in its scientific communities until the end of the thirties, the relation between the Italian Fascist regime and science seems to have been guided by the principle of effectiveness. Already in 1923 the Minister Gentile proposed, to the Italian Society for the Advancement of Science, that science had a moral responsibility to be undertaken in the service of the political and economic interests of the country. This approach, however, was not implemented until the thirties when autarky and military intervention required a re-examination of the need for national technological development. In this context, as Maiocchi shows in his chapter, after 1932 the CNR abandoned its secondary position to embark on the mission of realising the military–industrial potential of the Italian Empire. Since the mid-1930s, budgets multiplied and the Centre received significant

---

[11]   To assess the impact on science in other countries in terms of publications of physicists who emigrated see Cardona and Max (2005, pp. 313–34).

contributions, for the first time, from major Italian companies. In 1937 the CNR was freed from the Ministry of Education and reported directly to the Prime Minister's office (Maiocchi, 2004, pp. 27–50).

Obviously, the perceived utility of the products of science for national development was not overlooked by the Spanish and Argentinian military, despite their general reactionary mistrust of science. Ultra-nationalist values permeated the military who led both dictatorships and they aspired to scientific and technological development as the basis of economic development and, of course, military primacy. This kind of technocratic military existed in Argentina where the struggle against Chile and Brazil for military supremacy in the region was an important factor. As described by Jacovkis, these aspirations were held by the military (and engineer) President Agustin P. Justo, elected after a *coup d'état* in an election without the participation of the most popular political party, and also by General Manuel Savio. In Spain, despite the general reactionary attitude to science, military personnel (such as Juan Antonio Suances, the director of the state industries) were examples of people who held these technocratic attitudes; similarly José María Otero Navascués, whose later activity at the head of the Nuclear Energy Board is referred to in the chapter by Roqué. Significantly, given their authoritarian and anti-liberal characteristics, they all sank into the contradiction posed by Jacovkis: they craved the technological products of science, but they distrusted the scientific community.

Ideology played a very important role in all the dictatorships studied in this collection. However, the huge economic and military possibilities of science enhanced the logic that granted primacy to efficiency over ideology. For instance, as Maiocchi notes in his chapter, the search for efficiency meant that Mussolini protected the anarchist Molinari. Conflict between the two logics concerning science is illustrated by the Spanish case, in which the regime was forced to neglect its project of imperial science and to encourage the development of research for autarky. Canales argues in his chapter that technical efficiency was the way in which the Franco Regime reinserted itself into the international context during the Cold War, although this did not mean that the resultant science and technology escaped the marks of Francoism (Camprubí, 2014, Ch. 4). This process was more complicated in Argentina because, as Maffía explains, in the 1970s certain sectors of science held different broad conceptions of science that highlighted precisely the opposite values, that is, the social and ideological commitments of science that manifested in the debate between 'scientifism and anti-scientifism'. Anyway, as can be seen in Jacovkis's chapter, the search for efficiency resulted in many scientists purged at the universities finding refuge at the National Agency for Atomic Energy.

The balance between ideology and efficiency was not static and, common to all dictatorships, it shifted over time. In the early stages of the regimes, ideology generally dominated; effectiveness was emphasised later, both in war and autarkic contexts, when regimes needed practical results. This was the case for Francoism, but also Nazi Germany, where political and ideological hostility was soon counteracted by the search for efficiency (Kelly, 1985; Grüttner, 2005; Cornwell, 2003; Gómez, 2009), especially when the State was becoming the primary engine of scientific and technological development.[12] This was a gradual process that accelerated in the years of territorial expansion and during the Second World War, to the extent that the need of the Third Reich to obtain practical results became acute. In fact, Nazi authorities were soon concerned about the consequences of the primacy of the political–ideological criteria over those of quality, as were the Wehrmacht and industry. Authors such as Horn (1976) argue that the primacy of the political and the ideological led to a collapse of traditional technical and scientific quality criteria in the German educational system, which alarmed various ministries, the army and the SS. The principle of effectiveness figured prominently in making decisions about what research should be maintained and what to discontinue. When scientific projects did not meet expectations, they were questioned, even if they were ideologically in line with the regime. This was the case of glacial cosmology (Glazial-Kosmogonie) developed by SS leader Heinrich Himmler (Bowen, 1993), or the Aryan physics of Stark and Lenard (Walker, 1995).

The Relationship of Scientists with Politics

This book shows that the relationship between science and politics is not unidirectional.[13] Up to this point we have dealt with the intervention of politics into science, but the scientific community did not play a passive role in the relationship between science and politics.

Gómez's chapter illustrates the modulation of these relations within liberal-democratic Spain using the notion of the social contract for science. Scientists expected good working conditions, mainly resources to operate freely according to their principles. But, as Gómez shows, even in a liberal-democratic context

---

[12]   In Germany this extended to industrial science as the State appropriated the research of such industries as the IG Farben chemical conglomerate (Hayes, 2000).

[13]   And, as mentioned in the preface, the conceptual separation of science and politics as two realms that 'impact' on each other, although a useful enough heuristic in the present context of thinking about dictatorship and democracy, has limitations.

things were not so simple and easy, because political power intervened in the government of science and, on the other hand, scientific communities developed practices that called into question that they always acted in a disinterested manner.

Under Dictatorships, the relationships between scientific communities and politics changed in a remarkable way since political intervention became overt, direct and strong thus affecting the conditions for scientific practice. Scientists adapted to the dictatorial situation in different ways, as described in the various chapters of this book. Strategies varied: defensive isolation around professional values, resistance to some policies, collaboration while maintaining a certain independence, not to mention the extreme positions of commitment and opposition to the regime. Among them, the most common attitude of the scientific communities seems to have been the development of a wide range of pragmatic strategies, ranging from inhibition to accommodation through to gaining benefit.

Pragmatic inhibition was a strategy widely followed by scientists with regard to purges. It is not easier to find a clearer case of pragmatism than the collective oath of Italian scholars. As mentioned, 99 per cent of them decided, for different reasons, to bow before Fascist demands (Goetz, 2000, pp. 13–16). Pragmatism was even greater when purges affected only a portion of the scientists. The Italian scientific community remained silent about the growing anti-Semitic atmosphere, while it was still not the official position of the regime, and later maintained this silence towards the application of racial laws of 1938. Some authors, like Finzi (1997, p. 28), argue that after these laws the former silence became an active consensus.

The situation was similar in Germany, where non-Jewish scientists did not protest when their colleagues were purged and, in turn, they attempted to maintain the traditional rules that had governed the university and research institutes.[14] The priority for the management of the Kaiser Wilhelm Society was not to withstand the effects of the purifying process, but to control as much as possible so as to preserve its autonomy and avoid disturbances such as anti-Semitic demonstrations which accompanied the purge at the Medical Research Institute of Heidelberg (Beyler, Kojevnikov and Wang, 2005, pp. 26–8). This strategy of pragmatic inhibition had few exceptions even for eminent scientists like Fritz Haber, who despite his radical nationalism and imperial enthusiasm was Jewish. However, Max Planck's intervention with Hitler on his behalf resulted in the answer that all Jews were the same: communists and enemies

---

[14]   We must remember that there was a strong tradition of autonomy of science from politics in Germany.

(Noakes, 1993, p. 379). Similar situations occurred in other institutes such as the Psychological Institute of Berlin, which was led by Wolfgang Köhler who reacted ambiguously to the first wave of dismissals of Jews (Cornwell, 2003).

On the contrary, this pragmatic inhibition was not possible in Spain, since the purge was systematic and universal. All public employees, including scientists and scholars, were made redundant and had to apply individually for readmission through an inquisitorial process where the best way to defend oneself was to accuse others (Riquer, 2010, pp. 144–50; Morente, 2005). In this context, the space for silence in the face of determining other peoples' fates was minimal.

At the other extreme to inhibition, some scientists chose to get resources through strategies that we could consider very close to scientific fraud. This was the case for the Italian CNR when it offered Mussolini the results that the dictator expected. It was a deal that, paradoxically, satisfied both parties (Mussolini heard what he hoped to hear, and the CNR increased their resources), although it was ultimately detrimental for Italy in the war. As indicated, liberal and democratic contexts are not free from these situations. It is a truism that power, whatever its orientation, can be tempted to listen to what fits its needs and that unfortunately there is no shortage of scientific sirens willing to issue such songs. However, freedom and critique allow alternative and critical diagnostics to be available, possibly in less prestigious or accessible fora, but after all available. This is not the situation under dictatorships. Scientists act in a context where they may be fearful to criticise or disagree with what the powerful leader wants to hear. Personality-centred authoritarianism makes this perverse dynamic under dictatorships structural.

Different from fraud, but in complex connection with it, another dimension of the relationship between scientists and politics is the use of politics inside the scientific community: what we might call the *politician scientist*. In both the democratic and the authoritarian state scientists faced the temptation of using the power of the government not only to obtain resources, but also to defend positions of power and authority over other members of the community. Dictatorial contexts seem to favour this imposition of political and ideological criteria over scientific or academic ones. A spectacular example of the pre-eminence of politics to achieve power inside the scientific community is the former high school teacher and new head of Spanish scientific research by political appointment, José María de Albareda (Gómez and Canales, 2009), who from his position of power decided which scientific schools should be supported and which schools should disappear. As Huertas notes in his chapter, Albareda's corporate interests were behind the dismantling of the prestigious neuroscience school of Ramón y Cajal and the development of research lines that had nothing to do with neuroscience inside the research institute that kept the same name. Albareda aimed to control and enhance his own research area thus creating a new scientific field that so far had been non-existent in Spain.

The case of the rivalry between the two Italian statistical societies, described in this collection by Prévost, is another example of this use of State power inside the scientific community. In order to prevail in the face of scientific and academic rivalry with a figure as recognised as Gini, the society inspired by Livi sought powerful connections like Alberto De Stefani, a member of the Fascist Grand Council and former Finance minister (1922–25). In addition, leaders of this society participated in official bodies with undisputed Fascist content, such as the Higher Council of Demography and Race, an advisory body created after the enactment of racial laws.

Nevertheless, the relationship between scientists and politics goes beyond these pragmatic strategies aimed to get resources or power or to limit the effect of political intervention. Scientists – like everyone – also have ideologies as either professional or political commitments or as part of their wider world-views, and these ideologies manifest themselves in at least two important dimensions. The first one has to do with the relationship between scientific activity and society and the second with the beliefs, values and political options of scientists. As is shown by Gómez, Spain is a clear example of how scientists kept clear ideological positions behind their political involvement in science and technology and their effort to establish the conditions for improving scientific education and research in Spain. Since the nineteenth century, and certainly within the framework of the Board for Advanced Studies and the National Foundation for Scientific Research and Trials of Reform, different Spanish scientists participated in general politics (for example in the democratic revolution of 1868–74 or in the Second Republic), but prominently participated in science policy in various ways. In fact, Gómez demonstrates the involvement of Spanish scientists in science policy and their clear advocacy of the need to improve training and scientific research in the country. Ideas such as social responsibility, patriotism, the living conditions of the people, responsibility to improve the security and defence of the country or the economy were widely shared by Spanish scientists during this period.

An interesting dimension of the ideological and political commitment of scientific communities is nationalism. Scientific development of countries always involves a certain nationalist dimension even in democracies. This nationalist dimension was a key element in order to understand any scientific communities' accommodation to dictatorships.

Italy is a good example of the development of scientific nationalism that entered the service of the goals of the regime. In the context of frustration after the First World War, a discourse emerged that aimed to overcome the scientific backwardness of Italy and its dependence on technology through applied research. This approach was reborn in the 1930s with the recognition of the CNR, as discussed in Maiocchi's chapter. Great scientists who had remained suspicious of the regime embraced this cause in the second half of the 1930s: Giovanni Battista Bonino, the reputable Italian chemist-theorist, is one example. Many scientists saw science and technology, and scientific rationality, as the cornerstone of the new Italy and as the basis of the reconstruction of society. Even a purged and imprisoned man such as anarchist engineer Henry Molinari was seduced by this project (Maiocchi, 1993). Another example of the connection between scientific and nationalist approaches is the case of Gini, described by Prévost. Gini simultaneously expressed theoretical opposition in a debate about statistics alongside an expression of the supremacy of Italian statistics against Northern Anglo-Saxon powers.

The idea of a national science, subject to the state's objectives, was easily accepted. This idea was not new, given strong traditions of involving science and technology in national objectives such as industrial, economic or military development. Engineers and scientists (as happened in Italy for example with Gini or in Spain with Albareda, among others) believed they were working to a highly desirable end. A sense of service to the nation and a desire to contribute to a glorious future were key in supporting Fascists and their programmes. However, an interesting question is how to explain the involvement of scientists when this commitment to the national community provided the qualitative jump toward imperial aggression and dominance, and eventually toward mass murder of populations, as happened in the Nazi case. Highly reputable scientists and technicians played an important role in the East General Plan of 1942–43 that sought to establish German settlements in Eastern Europe and that involved mass deportation, slave labour and murder of the Jewish population. Agricultural experts, architects, demographers, geographers and geopolitical scientists participated in this project. Among them were the supporters of the regime, but also those who, despite their lack of sympathy for Nazi ideology, shared German imperialist purposes and had a clear vocation of service to their nation (Szöllosösi-Janze, 2001, p. 12). Scientists and engineers thought they had an important role in the construction of the new Germany, or new Italy, and therefore for the war effort.

Far from political commitment, a common self-perception of individual scientists working under dictatorship was paradoxically that their work had nothing to do with politics. In fact, many scientists that participated in the national scientific projects of their dictatorial governments maintained an attitude that sought to take refuge in neutrality and objectivity as the overarching values that governed their professional practice. As such, they thought of themselves as being limited to the investigation of specific problems 'scientifically', in physics, genetics, chemistry or any other field. Another different question was: to which (or whose) objectives should the results of their work be applied? Belief in the neutrality of the science, that scientific and technological rationality is not in charge of the ends but only of the means, and that science is wholly separate from its uses allowed these scientists ostensibly to distance themselves from politics and ideology while collaborating in the dictatorship's projects.[15] This mechanism, apparently derived from the essential nature of science, allowed them to inhibit all responsibility in the moral or political sphere.[16]

Conclusion

In this Introduction we have set out a general framework for understanding the particular historical narratives contained in the following chapters. We argued that during the nineteenth century the governments of Argentina, Spain and Italy showed an interest in scientific development as a source of enrichment and empowerment for their countries. At the beginning of the twentieth century, Spain and Italy put into practice science policies to generate scientific and technological development. Doing that, their actions coincided with the general trend of increasing state intervention in scientific activity that was

---

[15]   This way of thinking was so widespread that when Stark, the defender of the 'Aryan physics', requested that the twelve German Nobel laureates sign a letter of support for the Führer, Heisenberg's (and others) response was to claim that he 'personally agreed with the text but considered [it] unbecoming of scientists to make political statements' (Walker, 1995, p. 25). As noted by Szöllosösi-Janze (2001, p. 6), physicists, even during the war, felt that they were doing good physics and disallowed the intrusion of personal issues that might stain the purity of scientific knowledge. Once again, we add the caveat that the appeal to the political neutrality of science is not unique to scientists in dictatorships, as exemplified by Robert Oppenhemier's famous quote about the Manhattan Project and atomic bomb that: 'When you see something that is technically sweet, you go ahead and do it and argue about what to do about it only after you've had your technical success' (see Thorpe, 2006 pp. 223–8 for a discussion).

[16]   For an interesting treatment of scientific responsibility and the ethical dimension of science and technology, see Agazzi (2004).

developing in Northern countries such as Great Britain, Germany and France. The Argentinian governments maintained an ambiguous relationship with the scientific development of their country: this development was well supported by some governments but not by others.

Dictatorships provided a new context for scientific and technological development that broke with the liberal, constitutional or democratic framework in which science had recently developed. The attitude of the Dictatorships was different in the three countries considered in this book. The fundamentalist Catholic component of the Spanish and Argentinian dictatorships motivated an attitude of distrust toward science and, mainly, towards scientists and the ideal values of their community. Italian Fascism, on the contrary, encouraged scientific development and established an instrumental relationship with science and technology that sought to place them at the service of autarky and the war.

Consistent with these different attitudes, political and ideological intervention in the various scientific communities was much greater in Spain and in Argentina than in Italy. In Spain, the winners in the Civil War developed a systematic ideological purging which, in addition to exile, dismantled pre-war scientific communities. In Argentina, dismissal and exile of scientists and university teachers was common during different times, including some periods of constitutional government. In Italy, on the contrary, the continuity and the autonomy of scientific communities were greater, because their purge was a relatively late phenomenon and mainly racially motivated. Nevertheless, rising above the primacy of political and ideological principles, the logic of pragmatic effectiveness ended up being imposed. In Italy this logic prevailed from the beginning of Fascism, but even in Francoist Spain distrust toward science was left behind in the search for beneficial results.

Some scientists, on the other hand, developed strategies that allowed them to obtain resources under new conditions. Some who traditionally had defended the necessity of a science policy to develop the country found, in nationalism, a way to accommodate to dictatorial regimes. Others actively shared the ideological principles of dictatorships. Nevertheless, many scientists tended to take refuge in the traditional values of scientific neutrality to establish a strict separation between the scientific research that they developed and its use. This mechanism, ostensibly derived from inherent scientific standards, allowed them to avoid all responsibility in the moral or political sphere. These seemingly prototypical values of the scientific communities played a role as important in the technological and scientific production of these regimes as the more overtly political ideological aims of scientists.

List of References

Agar, J., 2012. *Science in the 20th Century and Beyond*. Cambridge: Polity.

Agazzi, E., 2004. *Right, Wrong and Science. The Ethical Dimensions of the Techno-Scientific Enterprise*. Amsterdam: Rodopi.

Balmer, B., 2012. *Secrecy and Science: A Historical Sociology of Biological and Chemical Warfare*. Farnham: Ashgate.

Benzi, M., 2010. 'Science and Fascism. Scientific Research under a Totalitarian Regime' [pdf]. Italian Studies Program and Department of French and Italian, Emory University. Available at: <http://www5.in.tum.de/~huckle/fasc.pdf> [Accessed 9 November 2013].

Bernal, J.D., 1939. *The Social Function of Science*. Cambridge, MA: MIT Press, 1967.

Beyler, R., Kojevnikov, A. and Wang, J., 2005. 'Purges in Comparative Perspective: Rules for Exclusion and Inclusion in the Scientific Community under Political Pressure'. In: C. Sachse and M. Walker (eds), 2005. *Politics and Science in Wartime: Comparative International Perspectives on the Kaiser Wilhelm Institute*. Chicago: University of Chicago Press, pp. 23–48.

Boatti, G., 2010. *Preferei di no*. Torino: Einaudi.

Bowen, R., 1993. *Universal Ice: Science and Ideology in the Nazi State*. London: Belhaven Press.

Bush, V., 1945. *Science – The Endless Frontier: A Report to the President on a Program for Postwar Scientific Research*. Washington: US Government Printing Office. .

Calver, N., 2012. 'The Royal Society and the Rothschild "Controversy" 1971–1972'. Available at http://royalsociety.org/events/2013/rothschild-controversy [Accessed 4 February 2014].

Camprubí, L., 2014. *Engineers and the Making of the Francoist Regime*. Cambridge, MA: The MIT Press.

Cannon, S.F., 1978. *Science in Culture: The Early Victorian Period*. New York: Science History Publications.

Cardona, M. and Max, W., 2005. 'The disaster of the Nazi-power in science as reflected by some leading journals and scientists in physics. A bibliometric study'. *Scientometrics*, 64(3), pp. 313–34.

Clarke, S., 2010. 'Pure Science with a Practical Aim. The Meanings of Fundamental Research in Britain, 1916–1950'. *Isis*, 101(2), pp. 285–311.

Coopey, R., 2002. 'Cold War, Hot Science: Applied Research in Britain's Defence Laboratories 1945–1990'. *British Journal for the History of Science*, 35(4), pp. 475–85.

Cornwell, J., 2003. *Hitler's Scientists: Science, War and the Devil's Pact.* London: Viking.

Dixon, T., 2008. *Science and Religion: A Very Short Introduction.* Oxford: Oxford University Press.

Edgerton, D., 2005. 'Science and the nation: towards new histories of twentieth-century Britain'. *Historical Research*, 78(199), pp. 97–112.

Edgerton, D., 2006. *Warfare State: Britain 1920–1970.* Cambridge: Cambridge University Press.

Edgerton, D., 2009. 'The "Haldane Principle" and other invented traditions in science policy'. *History and Policy* web-site, http://www.historyandpolicy. org/papers/policy-paper-88.html_[Accessed 27 March 2014].

Finzi, R., 1997. *L'università italiana e le leggi antiebraiche.* Roma: Editori Reuniti.

Fjæstad, M., 2010. *Research Institutes in Germany: Basic and Applied Science Institutionalized?* Department of History of Science and Technology, KTH, and Max Planck Institute for the History of Science, Berlin 2010. Paper No. 232. Available at http://www.kth.se/dokument/itm/cesis/CESISWP232. pdf [Accessed 3 March 2014].

Forman, P., 1987. 'Behind Quantum Electronics: National security as basis for physical research in the United States, 1940–1960'. *Historical Studies in the Physical and Biological Science*, 18(1) pp. 149–229.

Fox, R., 2010. 'La ciencia, el Estado y el bien público en Francia (1900–1940)'. In: J.M. Sánchez Ron and J. García-Velasco (eds), 2010. *100 años de la JAE. La Junta para la Ampliación de Estudios e Investigaciones Científicas en su Centenario.* Madrid: Residencia de Estudiantes, pp. 423–5.

Goetz, H., 2000. *Il giuramento rifiutato. I docenti universitari e il regime fascista.* Milano: La Nuova Italia.

Gómez, A., 2009. 'Ciencia y pseudociencia en los regímenes fascistas'. In: A. Gómez and A.F. Canales (eds), 2009. *Ciencia y fascismo: la ciencia española de postguerra.* Barcelona: Laertes, pp. 13–47.

Gómez Rodríguez, A., 2014. 'Frontera e integridad en el contrato social para la ciencia española, 1907–1939'. *Dymanis*, 34(2), pp. 415–87.

Gómez, A. and Canales, A.F., 2009. 'The rebels and the new Spanish scientific culture'. *Journal of War and Culture Studies*, 2(3), pp. 321–3.

Goudsmit, S., 1947. *Alsos.* New York: Schuman.

Graham, L.R., 2004. *Science in Russia and the Soviet Union. A Short History.* Cambridge: Cambridge University Press.

Grüttner, M., 2005. 'German Universities under the Swastika'. In: J. Connelly and M. Grüttner (eds), 2005. *Universities under Dictatorship.* Pennsylvania: Pennsylvania University State Press, pp. 77–111.

Gummett, P., 1980. *Scientists in Whitehall.* Manchester: Manchester University Press.

Guston, H.D., 2000. *Between Politics and Science. Assuring the Integrity and Productivity of Research.* Cambridge: Cambridge University Press.

Guston, H.D. and Keniston, K. (eds), 1994. *The Fragile Contract. University Science and Federal Government.* Cambridge, MA: The MIT Press.

Harwood, J., 1993. *Styles of Scientific Thought: The German Genetics Community, 1900–1933.* Chicago: University of Chicago Press.

Hayes, P., 2000. *Industry and Ideology: IG Farben in the Nazi Era.* Second Edition. Cambridge: Cambridge University Press.

Horn, D., 1976. 'The Hitler Youth and Educational Decline in the Third Reich'. *History of Education Quarterly*, 16, pp. 425–47.

Israel, G. and Nastasi, P., 1998. *Scienza e razza nell' Italia fascista.* Bolonia: II Mulino.

Jacovkis, P., 2003. 'Ciencia y política en Argentina'. *VI Congreso Nacional de Ciencia Política.* Available at: http://www.reflexionespys.org.ar/index. php?option=com_content&view=article&id=141:ciencia-y-politica-en-ar gentina&catid=29:diciembre-2003 [Accessed 7 November 2014].

Kelly, R.C., 1985. 'German Professoriate under Nazism: A Failure of Totalitarianism Aspirations'. *History of Education Quarterly*, 25(3), pp. 261–80.

Kitcher, P., 1993. *The Advancement of Science: Science without Legend, Objectivity without Illusions.* New York, Oxford: Oxford University Press.

Kitcher, P., 2011. *Science in a Democratic Society.* New York: Prometheus Books.

Macrakis, K., 1993. *Surviving the Swastika, Scientific Research in Nazi Germany.* Oxford: Oxford University Press.

Maiocchi, R., 1993. 'Scienziati italiani e scienza nazionale (1919–1939)'. In: S. Soldani and G. Turi (eds), 1993. *Fare gli italiani. Scuola e cultura nell'Italia contemporanea. II. Una società di massa.* Bolonia: Il Mulino, pp. 41–86.

Maiocchi, R., 2004. *Scienza e fascismo.* Roma: Carocci.

Merton, R.K., 1973. 'The Normative Structure of Science'. In: N. Storer (ed.). *The Sociology of Science.* Chicago: University of Chicago Press, pp. 267–78.

Morente, F., 2005. 'La universidad fascista y la universidad franquista en perspectiva comparada'. *Cuadernos del Instituto Antonio de Nebrija*, 8, pp. 190–96.

Morente, F., 2015. 'Entre tinieblas. La universidad española en la larga posguerra'. In: A.F. Canales and A. Gómez (eds), 2015. *La larga noche de la educación española. El sistema educativo español en la posguerra.* Madrid: Biblioteca Nueva, pp. 183–217.

Mulkay. M., 1979. *Science and the Sociology of Knowledge.* London, Boston: G. Allen & Unwin.

Nadal, A., 1987. 'Experiencias psíquicas sobre mujeres marxistas malagueñas. Málaga 1939'. *Baetica*, 10, 1987, pp. 365–84.

Noakes, J., 1993. 'The Ivory Tower Under Siege: German universities in the Third Reich'. *Journal of European Studies*, 23(4), pp. 371–407.

Pancaldi, G., 1993. 'Vito Volterra: Cosmopolitan Ideals and Nationality in the Italian Scientific Community between the *Belle époque* and the First World War'. *Minerva*, 31, pp. 21–37.

Picard, J.F. and Pradoura, E., 1988. 'La longue marche vers le CNRS (1901–1945)'. *Cahiers pour l'histoire du CNRS*, 1, pp. 7–40.

Polanyi, M., 1962. "The Republic of Science: Its Political and Economic Theory'. *Minerva* 1(1), pp. 54–74.

Polanyi, M., 1969. 'Knowing and Being'. In: M. Grene (ed.). *Knowing and Being: Essays by Michel Polanyi.* London: Routledge & Kegan Paul.

Popper, K., 1945. *The Open Society and Its Enemies.* London: Routledge.

Preston, P., 2008. *El gran manipulador.* Barcelona: S.A. Ediciones B.

Proctor, R.N., 1988. *Racial Hygiene: Medicine Under the Nazis.* Cambridge, MA: Harvard University Press.

Prost, A., 1988. 'Les origines de la politique de la recherche en France (1938–1958)'. *Cahiers pour l'histoire du CNRS*, l, pp. 1–18.

Richards, M., 2012. 'Antonio Vallejo Nágera: Heritage, psychiatry and war'. In: A. Quiroga and Arco, M. (eds), 2012. *Right-Wing Spain in the Civil War Era: Soldiers of God and Apostles of the Fatherland, 1914–1945.* New York: Continuum Publishing, pp. 195–224.

Riquer, B. de., 2010. *La Dictadura de Franco.* Barcelona-Madrid: Crítica-Marcial Pons.

Saraiva, T. and Wise, M.N., 2010. 'Autarky/Autarchy: Genetics, Food Production, and the Building of Fascism'. *Historical Studies in the Natural Sciences*, 40(4), pp. 419–28.

Szöllosösi-Janze, M., 2001. 'National Socialism and the Sciences: Reflections, Conclusions and Historical Perspectives'. In: M. Szöllosösi-Janze (ed.), 2001. *Science in the Third Reich.* Nueva York: Berg, pp. 1–37.

Thorpe, C., 2006. *Oppenheimer: The Tragic Intellect.* Chicago: University of Chicago Press.

Vallejo Nágera, A., 1938. 'Biopsiquismo del Fanatismo Marxista'. *Revista Española de Medicina y Cirugía de Guerra*, 3, pp. 189–95.

Vallejo Nágera, A., 1939a. 'Psiquismo del fanatismo marxista Investigaciones biopsíquicas en prisioneros internacionales'. *Revista Médica Española de Medicina y Cirugía de Guerra*, 11, pp. 53–8.

Vallejo Nágera, A., 1939b. 'Psiquismo del fanatismo marxista Investigaciones biopsíquicas en prisioneros internacionales'. *Revista Médica Española de Medicina y Cirugía de Guerra*, 12, pp. 132–43.

Vallejo Nágera, A., 1939c. 'Psiquismo del fanatismo marxista Investigaciones biopsíquicas en prisioneros internacionales'. *Revista Médica Española de Medicina y Cirugía de Guerra*, 14, pp. 299–308.

Vallejo Nágera, A. and Martínez, E.M., 1939. 'Psiquismo del fanatismo marxista. Investigaciones psicológicas en marxistas femeninos delincuentes'. *Revista Española de Medicina y Cirugía de Guerra*, 9, pp. 398–413.

Vinyes, R., 2001. 'Construyendo a Caín. Diagnosis y terapia del disidente: las investigaciones psiquiátricas militares de Antonio Vallejo Nágera con presas y presos políticos'. *Ayer*, 44, 2001, pp. 227–50.

Vittoria, A., 1991. 'L'Università italiana durante il regime fascista: controllo governativo e attività antifascista'. In: J.J. Carreras and M.A. Ruiz (eds), 1991. *La universidad española bajo el régimen de Franco (1939–1975)*. Zaragoza: Institución Fernando el Católico, pp. 29–62.

Walker, M., 1995. *Nazi Science, Myth, Truth, and the German Atomic Bomb*. Cambridge: Perseus Publishing.

Walker, M., 2001. 'Science and Ideology'. In: M. Walker (ed.), 2001. *Science and Ideology. A comparative history*. London: Routledge.

Withers, Ch., Higgitt, R. and Finnegan, D., 2008. 'Historical geographies of provincial science: themes in the setting and reception of the British Association for the Advancement of Science in Britain and Ireland, 1831–c.1939'. *The British Journal for the History of Science*, 41(3), pp. 385–415.

Wolfe, A.J., 2012. *Competing with the Soviets: Science, Technology, and the State in Cold War America*. Baltimore: Johns Hopkins University Press.

Chapter 2

The 'Social Contract' for Spanish Science before the Civil War[1]

Amparo Gómez

Introduction

The social contract for science is a model that has been developed to study the association established between science and politics in the twentieth century, through an analogy with the 'social contract' used in political theory (Mukerji, 1989; Kleinman, 1995; Guston, 2000; Guston and Keniston, 1994).The contract is the result of an acknowledgement by governments of the importance of science and technology for their country's development, and by scientists of the importance of public support for scientific research. The contract implies that government supports science and, in exchange, it is expected to obtain results which, although not specified in advance, are beneficial to the country. In other words, the basic idea is that governments invest in progress.

The social contract for science entails two general assumptions: a) the existence of a clear boundary between science and politics, based on independence and freedom of scientists to produce and manage science (the central assumption is that politics should not interfere in science);[2]and b) the integrity and productivity of scientists that guarantees the production of good science. The foundations of these assumptions are in the scientific standards that constitute an *episteme*, but also a *morality of science* (integrity, disinterestedness,

[1]   This chapter has been written thanks to the support of the Research Project FFI2009–09483 from Ministry of Science and Innovation and the Research Project FFI2012–33998 from Ministry of Economy and Competitiveness.
[2]   An example of such assumption would be the 'Haldane principle', which granted independence between science and policy. This principle was applied in the United Kingdom from the early twentieth century, and it was named in honor of Richard Burdon Haldane, who from 1909 to 1918 recommended this science policy.

universality or responsibility).[3] Therefore, in this model, all that is needed by science in order to develop and flourish is a great deal of freedom and sufficient resources.

Other assumptions are made in the application of the social contract for science to the analysis of history of science. Firstly, there is the idea that the relationships between science and politics characteristic of the classic social contract for science were held in Western countries, especially after the Second World War (Bush, 1945; Price, 1954; Steelman, 1947; Crowther, 1949; Crowther, Howart and Riley, 1942).[4] However, to the extent that science attracted the interest of governments long before, it is possible to find precedents for the contract since at least the late nineteenth century, and genuine manifestations of the contract after the First World War.[5] Secondly, it is considered that from the 1970s, given the changing nature of recent science, the assumptions of the social contract ceased to be always, or completely, satisfied (Guston, 2000; Bayles, 1983, Mulkay, 1973). Boundary problems generated by attempts to gain political control over science appeared, as did some integrity problems when scientific communities failed to act in accordance with the aforementioned epistemic and moral standards. These facts have resulted in a review of the model, which resulted in a more sophisticated version: a *new social contract for science* (Gibbons, 1999; Lubchenco, 1998; Demeritt, 2000; Slaughter and Rhoades, 2005; Pielke, 2007; and Gómez and Balmer, 2013).

Nevertheless, the classic social contract for science continues to be considered a useful analytical resource since it provides an interesting theoretical framework for clarifying certain political features concerning the history of scientific

---

[3]   These conditions are expressed by Merton's scientific *ethos* (communalism, universalism, disinterestedness and organised scepticism) (1974), or Polanyi's *dynamic orthodoxy* (mutual authority established between scientists, discussion, debate) (1962). Epistemic standards have been a central topic of twentieth century philosophy from logical empiricism to social and political epistemologies. Kleinman has argued: 'scientific practice was and should be independent of the rest of society and the scientific community should be responsible only to itself' (1995, p. 2).

[4]   It should be noted that in this period, especially during the Cold War, military research played an important role for governments, see for example, Balmer (2001, 2012), Edgerton (2006). See the analysis of international origins of scientific policy in Santesmases (2008, pp. 293–326).

[5]   Pielke (2007) analyses these relations in the USA between 1850 and 1940. He believes that genuine social contracts for science were established especially after the First World War. Guston (2000, pp. 3, 6 and 8) argues that the relationship between science and politics was clear in certain periods since the late nineteenth century, and scientific development was clearly the interest of politicians; see also Kleinman (1995, pp. 24–51).

development. The classic version of the contract might be useful to analyse early stages of scientific development which, in theory, would conform to the model. This is the case for the events that enabled scientific development in Spain in the early twentieth century. The social contract perspective would help to clarify the complex process of Spanish scientific development in this period. The contract bestows a certain degree of cohesion on studies of this process, shedding new light on this period of Spanish scientific development.

The analysis of Spanish scientific development from this perspective is based on certain research arguments, which will guide this study. These arguments are: a) that very early on, much of the Spanish scientific and intellectual elite envisaged a relationship between scientific development and the country's progress, thus triggering the demand for a science policy that was vital to the constitution of a social contract for Spanish science; b) that at the beginning of the twentieth century, a social contract for science was established in Spain, giving rise to several decades of flourishing science (and technology); c) that the 'social contract' for Spain's science policy was proposed by the liberals; and d) that this contract was subject to several boundary and integrity problems that were made explicit in the political debate.

The plan of this chapter is as follows. Firstly, it will examine the antecedents of the social contract for Spanish science, highlighting the early perception of the importance of science for the country's progress. Second it will analyse the constitution and characteristics of this contract, including the debate that was generated between conservatives and liberals about the politics of science that should be conducted in Spain. Finally it will be shown that the social contract for Spanish science was highly productive but also how the establishment of Franco's Regime after the Civil War (1936–39) marked the end of this contract.

The Background of the Social Contract for Spanish Science

Throughout the nineteenth century the Spanish scientific and intellectual elite gradually became aware of the importance of science and technology for the country's development and modernisation. This elite perceived how science and technology were playing a key role in the generation of wealth, the improvement of living conditions, and the economic and military predominance of certain European countries as Germany, England and France.

In contrast to these countries, there was little scientific activity in Spain and what existed was the result of isolated efforts of just a few eminent scientists. The most basic research resources were lacking and scientific–technical underdevelopment was considerable: 'scientific activity appears confined to sporadic efforts of a minority of disregarded men of science completely isolated from their social environment' (Núñez, 1975, p. 121). Manuel Revilla – a prestigious intellectual and one of the main authors of the 1845 Public Instruction Plan – pointed out that Spain would be developed 'when the manufacturing and industrial arts, to which science gives movement and life, become a new source of wealth and prosperity for families and towns', but in Spain 'the physical-mathematical and natural sciences are deemed to be of little importance [...], lacking entirely those material means which are vital to their fruitful study' (Revilla, 1854, p. 13).

In light of this situation, progressive intellectuals and scientists were in favour of modernising the country and promoting its development, opening it up to new models of thought coming from Europe. This ideal existed alongside traditionalist and conservative ideas in which religion and the Catholic Church carried great weight. Therefore, the key characteristic of Spanish society during this period was the complex balance between old and new ideas. This situation affected science also, and was clearly reflected in what has become known as 'the controversy of Spanish science'.

The Controversy of Spanish Science

The controversy of Spanish science, in fact, dates back to the eighteenth century, when an article entitled 'About Spain' was written by the French intellectual Masson de Morvilliers and was published in 1782 in the *Methodical Encyclopaedia*. The article claimed, among other things, that 'Spain may perhaps be the most ignorant nation in Europe' (Masson de Morvilliers, 1782, p. 51). The reason for this situation was the dominance of the Catholic Church, and the lack of freedom of thought resulting from the Inquisition's tight control over ideas.

The article prompted a number of defensive responses, such as those given by Antonio José Cavanilles (1784), a priest and naturalist, and the Abbe Carlo Denina (1786), an Italian historian who claimed Spain's historical contributions to metaphysics and theology. They defended the importance of these disciplines against modern science, since they were fields in which the Spanish had shone, and denied that natural science had anything to offer the country or its wellbeing. Spain, they claimed, should continue down the historical path of theology, the arts and military science which had contributed to its greatness. Fierce debate ensued with advocates of Masson de Morvilliers's article in Spain, whose writings appeared in the Madrilenian weekly newspaper *El Censor*, between 1785 and 1787 (Caso, 1989, p. 785).

The controversy re-emerged in the middle of the nineteenth century during a moment of renewed interest in science. The confrontation was triggered this time by a speech given by José Echegaray (an important mathematician and engineer) upon his official acceptance into the Royal Academy of Science in 1866. The speech was entitled *The history of pure mathematics in Spain* and, in addition to charting this history, Echegaray highlighted the lamentable situation of science in Spain. The speech was responded to by the mathematician and journalist Felipe Picatoste, who defended Spanish science and recalled 'the centuries in which we provided textbooks to the whole of Europe' (1866, p. 196). The debate continued on into the first decades of the twentieth century, with the participation of many respected scientists including José Rodríguez Carracido (1897, 1911), Santiago Ramón y Cajal (1897), Julio Rey Pastor (1915) and Gregorio Marañón (1941). All of these scientists underscored the lack of Spanish contributions to modern experimental science and pointed to Spain's isolation from the rest of Europe, together with the neglect of scientific research and scientific education, as the main reasons for this situation. The opposing position was defended by the Catholic traditionalist Marcelino Menéndez Pelayo (1894), as well as by other conservatives. Menéndez Pelayo stressed the importance of less prominent scientists in the history of science and argued that it was possible to find many Spanish examples of such scientists; he also highlighted the merit of Spanish philosophers.

The Krausists

One of the groups of liberal intellectuals who believed that science had a vital role in Spain's development, as well as in the development of countries and humanity in general, was the Krausists with members such as Julián Sanz del Río, Francisco Giner de los Ríos, Nicolás Salmerón and Manuel Cossío, among others. Krausism was an important intellectual movement in Spain with various political and social derivations. It was introduced mainly by Professor Julián Sanz del Río, who was fascinated with the thought of German idealist philosopher Karl Christian Friedrich Krause.[6] Sanz del Río travelled to Germany between

---

[6]   Krause's philosophy reconciles the idealism of Kant and Fichte with the idealism of Schelling and Hegel. His philosophy assumes that science – as *Wesenlehre* or *science of Being* – contains all knowledge, and the particular sciences (empirical sciences) are integrated inside the universal science or *science of Being* (Jiménez, 1985, pp. 42–3). The existence of God is proved through subjective analysis of the contents of consciousness of oneself. This is the *analytical method*: from the particular to the general, to the whole. But to analyse the objective reality of these ideas it is necessary to have categories derived down from whole to parts,

1843 and 1844 and adopted Krause's philosophy, considering this philosopher as the last great philosopher of German idealism. The reasons for his choice were the 'Spiritual and philosophical affinities and ethical and practical implications for social and political reform of Spain' (Jiménez, 1985, p. 97). He considered Krause's philosophy as 'true in theory and fecund in practice' (Sanz del Río, 1860, pp. XLI–XLII). This practical dimension was important for him since he was interested in the moral and intellectual modernisation of Spain. He did not limit himself to the introduction of Krause's thought in Spain, but adapted his works to his own purposes: to modernise Spain and raise the intellectual and moral level of the country. In fact all the Krausist intellectuals insisted that Spain needed to be modernised and the philosophy and practice that would allow it were inspired by Krause.

The Krausist programme for Spain's progress was based on three key ideas: the country needed to be educated, rationalised and moralised. Education was a key factor because it integrated science, rationality and moral doctrine. The Krausist defence of ideas such as 'academic freedom' and 'freedom of thought', along with their commitment to education independent from the State and the predominance of rationality, even over faith, were a vital part of the Spanish liberal scene from the mid-nineteenth century.[7]

Krausism had a major influence on the democratic period of 1868–74 (which included the First Spanish Republic). Krausist ideas and political intervention were at the heart of the democratic government's attempts to improve the education and science systems. One leading Krausist, Nicolás Salmerón, became Prime Minister of a government (of which Echegaray was also a member) that left university legislation and reform in the hands of the Krausists. They promoted the improvement of education and scientific research and teaching freedom, thus eliminating denominational teaching and severely restricting the influence

---

from the general to the particular, through the *synthetic method*. As noted by Tiberghien: 'Of Course, it can be concluded with certainty that God contains in itself Spirit, Nature and Humanity and that he is distinguished from each of these beings as the Supreme Being. God is immanent in all things' (Tiberghien 1875, p. 358). This is the essence of the Krausist position called *panentheism* that basically means not that the world is God (pantheism) but it is *in God* and is distinct from God. Krause questions the dualism claiming there is no antithesis in God, the physical world and the spiritual world. And spirit and nature are harmonised in Humanity. Nature and Spirit are conceived in the same plane and likewise its particular manifestations of body and soul in human beings. From this thought comes the respect and defence of rights of nature and the human body characteristic of Spanish Krausism. For these thinkers, also comes the defence of the rights of children, women and animals.

[7]   Krausism had an enormous influence on Spanish liberals (intellectuals, scientists and politicians), see Gil (1975).

of Catholicism. However, the Krausists recognised that the conditions were not right for applying the principle of independence between education and the State, since Spain was not yet ready for such a move, neither morally nor with regard to material resources. Furthermore, it was the Catholic Church that was best placed to set up and run private schools and universities, which meant that private education would still remain chiefly in its hands. Nevertheless, the Krausists continued to believe that the best way to improve the country's education was through private teaching institutions. They were convinced that education and scientific research in Spain would only develop to European standards once such institutions had been set up. In 1874 the democratic experiment, which had lasted for just six years, came to an end, and the monarchy was restored by a new conservative government. From this date to 1923, a political regime known as the *Restoration* was implemented. This regime was based on a pact made between the political elite to enable the peaceful alternation of moderate liberal and conservative governments under a non-democratic constitutional regime.[8] Under the new conservative government of 1875, academic freedom was rescinded. Teaching content was once again obliged to comply with the doctrines of official Catholicism and many other progressive aspects of university teaching were abolished. This generated a major conflict, known as the 'university question' with liberal lecturers, and several significant Krausist and liberal university professors were removed from their Chairs (Álvarez and Vázquez-Romero, 2005; Jiménez, 1985 and Cacho Viu, 1962).

In light of this situation, the Krausists decided to establish an independent, private university which would play a key role in promoting education and science in Spain. Thus, the Free Institution for Teaching (*Institución Libre de Enseñanza* – ILE) was founded by Giner de los Ríos in 1876, in accordance with Krausist ideas regarding the modernisation of the country through education, morality, reason and science. However, the new university never actually functioned as such, and the ILE focused more on primary and secondary education, implementing a novel, experimental pedagogic programme which constituted one of the most important antecedents of pedagogical reform in Spain. Important liberals and republicans were educated in the ILE and many of them gave financial support to this institution for years (since it was a private association). The influence of the ILE was palpable at the beginning of the twentieth century when some prominent Krausist intellectuals played a key role in the constitution of a social contract for Spanish science in 1907.

---

[8]   As Gil (1975, pp. 133–4) has pointed out: 'The Restoration ended with the social predicament that Krausism had enjoyed during the sexennium'.

Positive Mentality and Regenerationism

In addition to Krausism, another two schools that strongly influenced the renewal of Spanish thought at the end of the century were evolutionism and Comte's positivism (Núñez, 1975, 1977; Rodríguez Carracido, 1897, 1911, 1911a, b, and Glick, 1982). These schools had two major effects: a) they questioned Krausism, which lost influence as a system of ideas and was, as Hermengildo Giner stated, 'reduced to a flexible intellectual attitude, open to new scientific contributions' (Giner, 1912, p. VI) giving rise to krausopositivism;[9] and b) they introduced a positivistic mentality which trusted in science as the means of resolving all the country's problems. In the words of Núñez:

> one of the most fruitful and relevant characteristics of this positivist stage of contemporary Spanish thought is, no doubt, the creation of a widespread scientifistic-positive mentality, ranging from the toils of development, at the basic level of natural science [...], to approaches interested in the foundations of social and political action. (Núñez, 1975, p. 119)[10]

And Santiago Ramón y Cajal, future Nobel Prize-winner, stated: 'The books of Comte, Littré, Huxley, Darwin, Haeckel, Herbert Spencer [...] had brought about an incredible revolution. The splendour of science had dispelled the mists of metaphysics' (1923, p. 247).

Thus began one of the periods in Spanish history in which science and the values it embodied played a predominant role; a period in which confidence in the capacity of science to dispel obscurantism was combined with the conviction that it was the driving force behind the country's development. Science, it was thought, would show the way forward in the search for solutions for the diverse problems facing the country: economic development, industrialisation, education, relations between the centre and the peripheral regions, and many more such problems. In all cases, it was necessary to proceed according to science: observing and experimenting in order to establish the data, then analysing it impersonally and dispassionately, and inferring solutions exclusively from that data. As the Catalonian positivist, Pompeyo Gener stated: 'We do not

---

[9]   Positivists were critical of Krausism which they considered as Metaphysics, but also were influenced by Krausism giving rise in the late nineteenth century to *krausopositivism*. Prominent Krausists such as Nicolás Salmerón, Urbano González and Giner de los Ríos belonged to this movement (Díaz, 1973).

[10]   Anglo-Saxon science and philosophy had a great influence in Spain through the reception of evolutionism, organicism and utilitarianism.

form an opinion unless it is regarding that which we have personally observed or experienced' (Gener, 1877, p. 9); thus, even the Catalonian question should be dealt with by 'gathering as much data as possible in accordance with the modern scientific method' (Gener, 1903, p. 258). Therefore, for him: 'Politics should not be pure sentiment, but rather an inductive Science, like all sciences are today' (Gener, 1877, p. 325).

Religion, tradition, ideology and the old moral order were countered by objectivity, neutrality, dispassionate observation, precision and rigour. Independence of criteria, integrity, public open debate, reason and responsibility were defended in the face of webs of influence, privilege, corruption, pompous demagoguery and rhetoric. These values together made up a 'morality of science', the purest expression of which was scientific activity itself and scientists, who in this way began to acquire public and political relevance.

Science and its morality played an important role in the regenerationist movement for the modernisation of Spain.[11] Leading scientists such as Santiago Ramón y Cajal, Eduardo Hinojosa, Julian Rivera, Ricardo Macias Picaveda, José Rodríguez Carracido, José Echegaray, and the aforementioned Pompeyo Gener, took part in this movement.[12] The regenerationist movement had enormous influence at the end of the century, particularly from 1898 after the shock resulting from Spain's military defeat at the hands of the United States and the loss of the last colonies (Cuba, Puerto Rico and the Philippines). Science and technology were at the heart of all criticism regarding the outcome of the conflict and underpinned all arguments relating to the need to regenerate the country. As Ramón y Cajal pointed out:

> but more than anything, it was the shameful ignorance in which our parties lived in relation to the magnitude and true efficiency of both, their own and the other side's strength that dragged us towards the catastrophe. Because, although it may seem absurd, at that time, MPs, journalists, the military and so on, all believed in good faith that our military instruments in Cuba and the Philippines (wooden

---

[11]   *Regenerationism* was a movement that reacted to the 1898 colonial defeat. *Regenerationism* proposed a set of political, economic and social reforms with the intention to 'regenerate' the country, and asserted that this should be carried out with an objective and scientific mentality. For the role of *regenerationism* in the modernisation of Spanish education, see Canales (2013).

[12]   Cajal believed that the regenerationist thinking belonged to a minority movement not followed by the rest of the population and was disconnected from the masses: 'The regenerationists of 98 were only read by ourselves' (Ramón y Cajal, 1923, p. 268).

ships and an army of sick men) could overcome the formidable equipment
possessed by the enemy (Ramón y Cajal, 1923, p. 265).

And in his epilogue of 1899 he stated: 'To ignore that we were unaware of
the irresistible force of the adversary: the science of their engineers and their
chemicals (inventors of incendiaries that swept the deck of our ship and impeded
any defense), the superiority of their ships and the habergeons ...' (Ramón y
Cajal, 1941, pp. 219–20).

The deplorable condition of science and scientific training, even at the end of
the nineteenth century, is illustrated by the words of Rodríguez Carracido, who
recalled that when he took up his Chair in Organic Chemistry in 1899, 'we had
only a chair for the oral presentation of the lectures in biological chemistry, and
were totally lacking any working instruments, not only for students' practical
tasks, but also for testing the most simple phenomena explained during the
course of the lectures' (Rodríguez Carracido 1911a, p. 400). He claimed that
'for fourteen years, Biological Chemistry was explained as if it were Metaphysics!
[...] with all ministers (regardless of their political party) unanimously resisting
our demands for the resources necessary for the constitution of the laboratory'
(Rodríguez Carracido, 1911b, p. 389).

In this context, criticism of the state of Spanish scientific and technical
research became increasingly harsh, and leading regenerationist scientists
demanded the development of a science policy to put an end to this situation.
This demand for a science policy marks an important turning point in relation
to the previous stage, in which dissenting voices only focused on reporting
the situation of scientific backwardness. Now, however, regenerationists were
proposing true plans for political action, with clear objectives and goals.

The need for a science policy was clear for Ramón y Cajal, even before the
defeat of 1898. In his paper of 1897, *The Duties of the State in relation to scientific
production*, he stated that 'the posterity of nations is the work of science and its
application to life and to material interests'. Thus, 'it is the inescapable duty of
the State to stimulate and promote culture by developing a *science policy*' (Ramón
y Cajal, 1897, p. 373). This 'science policy' should be based on high-quality
scientific education, foreign study grants and the establishment of a network of
laboratories and research centres for scientific research in Spain. These measures
were partially similar to those proposed by the Krausists to Segismundo Moret's
liberal government in 1906, except in regard to the role that should be played
by universities. Thus the subsequent integration of regenerationist scientists,
including Ramón y Cajal, into the Krausist project was a fairly straightforward
matter.

Two Projects for the Development of Spanish Science

At the beginning of the twentieth century, both regenerationists and Krausists (and even some moderate conservatives) agreed on the need to develop a science policy to foster high-quality scientific research in Spain. However, this agreement did not extend to the specific policies through which the State should act. Ramón y Cajal, and the regenerationists, believed that any action taken should focus on the renewal of existing institutions, that is, public universities (a position shared by conservatives). For Cajal, science policy meant: 'Transforming the university, which until today has been almost exclusively dedicated to the granting of degrees and to professional teaching, into a place for intellectual impulse, as it is in Germany, where the University constitutes the principal body of philosophical, scientific and industrial production' (Ramón y Cajal, 1897, p. 374).

The social contract for science demanded by this science policy was institutional, it should be established between the government and the universities, open to all scientists either in or accessing the universities through established paths. Universities would be reformed through the implementation of several necessary measures, including an increase in funding and grants for studying abroad. These measures would result in modern science and technology (Ramón y Cajal, 1897, pp. 373–4).

However, the Krausists, with Giner de los Ríos and Cossío as leaders were suspicious of universities and believed that it was necessary to develop new scientific and training institutions outside the existing university system; such institutions should also be completely independent of the State (Cacho Viu, 1962, pp. 194–6). The aim was to modernise, rather than just regenerate, in accordance with the Krausist ideal of State independence. In this case, the contract would be between the government and scientists, based on an agreement whose key contents were, on the one hand, the self-government of science and almost complete independence from the State – whose role should be limited to financing – and on the other, the production of modern science.

In the end, the Krausist option prevailed, and the social contract for Spanish science resulted in the creation of two new scientific institutions which brought together most of the political and economic effort expended in support of Spanish scientific and technological restoration during the first third of the twentieth century. These institutions were: the Board for Advanced Studies and Scientific Research (*Junta para la Ampliación de Estudios e Investigaciones Científicas* – JAE), established in 1907, and the National Foundation for Scientific Research and Trials of Reform (*Fundación Nacional para Investigaciones Científicas y Ensayos de Reformas* – FNICER), established in 1931.

The political decision was to make scientific research independent from the universities and place the project in the hands of a group of scientists and intellectuals who would have a large degree of autonomy and capacity for self-government. The reasons for this political decision in favour of the Krausist proposal were complex, but basically it was due to the fact that the project was shared with politicians in the liberal government who had strong sympathy for Krausism.[13] The liberal Prime Minister, Segismundo Moret, and the Minister for Public Instruction and Fine Arts, Amalio Gimeno, had both had relationships with the ILE and the Krausists (Moret was a shareholder of the ILE and had even chaired some of their meetings). In fact, Segismundo Moret was an old friend of Giner, and in 1906 asked him for advice regarding how best to develop a policy for improving education and scientific research in Spain. Giner wrote a famous letter to him (dated 6 June 1906) in which he outlined the basic points of the policy he believed the government should implement (Castillejo, 1997, pp. 326–30). In the letter, Giner explained the aim of his project: to train new high-level staff as quickly and intensively as possible and to foster research by creating high-quality laboratories and centres. It was therefore necessary to provide the resources and ensure 'a considerable increase in both, number and size of foreign study grants, as well as more training in research and experimental teaching' (Castillejo, 1997, p. 327). Moreover, it was also necessary to develop centres which were independent from partisan action. All this should be done outside the universities, and the task should be entrusted to a small group of competent, enthusiastic people who represented the different currents of thought present in the country. It was vital to have well-trained people both in science and education, since this had been one of the historic problems of Spain's backwardness due to its scientific isolation from Europe. The opening up of Spain to Europe and European training were vital requisites for modernisation (for both Krausists and regenerationists).

High-quality scientific production was essential to the country's economic development and modernisation, and Moret's liberal government made a commitment to ensuring its development by founding an institution specifically to pursue this goal and providing it with both autonomy and resources. However,

---

[13]   Moret and Giner discussed the project at previous meetings. This demonstrates that 'official' decisions have a background of negotiations, agreements and debates that are not always reflected in the official documents; for a discussion see Balmer (2001, p. 5), 'documents frequently record much negotiation and dispute both before and after the "official" decisions'.

this commitment came about at the end of his government's term of office. The JAE started its pathway in a delicate political context in which a clearly liberal project was forced to begin under a conservative government which shared neither its aims nor its values.

The Social Contract for Spanish Science: the JAE

The Board for Advanced Studies and Scientific Research was established by Royal Decree on 11 January 1907 (in the last weeks of Segismundo Moret's liberal government).[14] This Decree indicated the functions and activities of the JAE, the way in which it was to be managed and its status as a public state-funded institution (dependent on the Ministry for Public Instruction and Fine Arts).[15] Article one outlines, among others, the following functions of the JAE: 'Extending studies both within and outside Spain', 'Delegations to scientific conferences' and 'The fostering of scientific research works' (GM, 15, 15 January 1907, p. 166). Two weeks later, on 27 January 1907, a set of internal regulations were drawn up and approved by the Plenary Board of the JAE. These regulations specified the independent operation of the JAE.

The importance of the project prompted the participation of Spain's leading scientists from a wide range of different political and intellectual fields, including

---

[14]   An earlier version of the role of the JAE in the making of the Spanish social contract for science has been discussed in Gómez (2014). For an approach to the classic studies of the JAE see, for example, Laporta, Solana, Ruíz, Zapatero and Rodríguez (1987, 1987a); Sánchez Ron (1988, 1998 and 1999); Sánchez Ron, Lafuente, Romero de Pablos and Sánchez (eds) (2007); Sánchez Ron (2010); Romero de Pablos and Santesmases (2008); and Otero and López (2012).

[15]   The JAE focused on pure science and paid hardly any attention to applied research. The second part of the social contract for science was carried out by the National Foundation for Scientific Research and Trials of Reform (FNICER) which was set up in 1931 at the beginning of the Second Republic (Royal Decree of 12 July 1931), and which only operated for four years before the outbreak of the Spanish Civil War. The FNICER was the only serious attempt made at supporting applied science and technology, and therefore science in relation to industrial and military development. It was made up of a considerable number of scientific-technical centres, institutes and laboratories. The aim was to connect research to private and industrial interests, and to this end scientific collaborative relationships were established with universities and technical schools. The FNICER was a decentralised organisation and right from the beginning was interested in cooperating with any relevant ministerial, municipal or citizen organisation. It was, therefore, a much more modern institution than the JAE, and much more in tune with the new era and its demand for applied science and technology. The problem was that it hardly had any time to develop its projects before the Civil War started.

Santiago Ramón y Cajal, José Echegaray, Luis Simarro, Ignacio Bolívar, José Rodríguez Carracido and Leonardo Torres Quevedo, among many others. Intellectuals such as Marcelino Menéndez Pelayo, Ramón Menéndez Pidal and José Castillejo were also involved in the JAE. Given the scientific status of Ramón y Cajal, he was proposed, from the beginning, as Chairman of the JAE (and indeed he served in that capacity until his death in 1934). He was succeeded by Ignacio Bolívar, who retained the post until the Board was dissolved in 1939 at the end of the Civil War. José Castillejo (one of Giner's closest disciples) was the secretary of the JAE until 1931 and the true manager and organiser of the whole project. Indeed, mainly through Castillejo, the Krausists Giner de los Ríos and Cossío exerted considerable influence on the JAE, even though technically they were never members.

The voting members and the secretary together formed the Plenary Board. The Executive Committee was formed by the Chairman, the two Deputy Chairmen, two members and the secretary. In 1907, the JAE's budget was 328,000 pesetas, an enormous sum considering the budgetary restrictions imposed at the time on the universities.

The same Decree of 1907 firmly established the independence of the JAE: 'The Board will have capacity to acquire own and manage property of all kinds destined for the purposes for which it is created, [...] will determine the distribution of grants, the procedure for granting them and requirements necessary to qualify for them' and also it will set the 'amount, duration and place of stay of grants' (GM, 15, 15 January 1907, p. 166). The JAE satisfied the objectives outlined by the Minister Amalio Gimeno in the preamble of the 1907 Decree: 'but in order for all this to be effective, this project needs to have a national character, being conducted in a perseverant, regular way by a neutral organization which, located outside the agitation of political passions, through all changes keeps its independence and prestige' (GM, 15, 15 January 1907, p. 166). And the Minister stated in the final paragraph: the key to transformation lay in ensuring that, throughout all ministerial changes, 'a few illustrious men' remained at the forefront of the endeavour, since it was now 'understood, as in other nations, that we must liberate this institution from administrative obstacles which, while generating apparent external equality, nevertheless excludes and replaces direct personal action with official action, which tends neither to be as fast nor as correct' (GM, 15, 15 January 1907, p. 166).

It was therefore the 'illustrious men' who were responsible for managing (in an independent manner from administration) the resources received, the grants, and the JAE centres and laboratories.[16] The basic premise was that scientific training and production of prestigious science required total freedom for scientists, in whose hands all decisions should be left. There was a clear boundary drawn between science and politics which no party should cross and which was guaranteed as long as both parties fulfilled their role. Scientific standards and the 'morality of science' assumed by both scientists and politicians ensured that this remained so. It was these convictions that made it possible to establish the type of social contract for science embodied by the JAE, which was ratified by successive liberal governments, as well as by the Republican government of the Second Republic but not by conservative governments.

The freedom granted to the JAE was total in relation to one of its principal functions: the awarding of postgraduate grants for studying abroad (and in Spain too). This was a key area of activity, for which a large percentage of the resources managed by the JAE was earmarked. In this respect, the articles of the 1907 Royal Decree were clear, article six states that: 'The Board shall determine the distribution of grants, the procedure by which they are awarded and eligibility requirements in accordance with article 5. It shall also determine, in accordance with the circumstances of each case, the amount and duration of the grant and the location in which study shall take place' (GM, 15, 15 January 1907, p. 166). Therefore, the grants were awarded in accordance with criteria established internally by the JAE itself, rather than on the basis of externally set criteria.

Another of the JAE's most pressing concerns was to ensure that young students returning from abroad found places in the Spanish academic world. This problem was partially solved by the JAE's capacity to hire those it believed should continue with their research and to establish what remuneration they received. This possibility was included in the internal regulations of 1907. Article 40 stipulated that: 'After the end of the grant period or after returning from abroad, the Board may require grantees not belonging to the teaching staff to remain for a certain time conducting research and teaching work in their particular area of interest, under the supervision of a Board representative and in return for remuneration proposed by the Board' (GM, 173, 22 June 1907, p. 1140). These young people were placed in the research centres and laboratories that were gradually established. The JAE, therefore, was completely free to decide who should remain researching and what salary they were to be paid;

---

16   As stated in the first JAE Report: 'The Board shall be the instigator of an intensive, fast renovation of our higher education and scientific research system, based on communication with foreign scientists, disinterested work and freedom of choice as regards subject matter and procedures' (JAE Report, 1907, p. 18). All Reports can be found in the official archives, on the JAE website: http://edaddeplata.org/tierrafirme_jae/memoriasJAE/index.html.

it also acted with complete independence in relation to the research centres it developed and the scientists appointed to run them.[17]

This extensive degree of autonomy granted to the JAE by liberals was a point of controversy from the year of its establishment and onwards, particularly under conservative governments.[18] Such autonomy was at the heart of a political debate in which the social contract for science embodied by the JAE and the liberal concept of science policy that underpinned it was called into question.

Boundary Conflicts

Shortly after the establishment of the JAE, the conservatives regained power. The new government was chaired by Antonio Maura, with Faustino Rodríguez San Pedro as the new Minister for Public Instruction and Fine Arts. During the three years of Maura's government, Rodríguez San Pedro put many obstacles in the path of the JAE project. The first move made by the new Minister was to modify the internal regulations of the JAE related to its autonomy. These modifications triggered a radical change in the nature and functions of the Board (JAE, 1908–1909). The principal modifications were: a) all Board agreements 'not just the budgets but also those relating to purely scientific matters, shall be submitted to the Minister for approval' (Castillejo, 1997, p. 350); this included 'technical decisions [...] and the resources other than those granted by the state' (Castillejo, 1997, p. 350); b) all JAE grants to study abroad had to be submitted to the Minister for approval; c) a shortlist of three names for both grants and new members had to be presented to the Minister, who reserved the right to make the final decision (in certain cases, the Minister could decide without a prior proposal from the JAE); and d) the JAE's budget was reduced (from 328,000 to 150,000 pesetas). Finally, in relation to the foreign study grants (hardly any of which were granted that particular year) the minister reminded the JAE that the money for the grants depended on political decisions.

The situation remained unchanged for the remaining three years of the conservative government's term of office. With the decrees of 9 June 1908, 10 July 1909 and 29 January 1909, the conservatives established even stricter

---

[17]   For this topic see González and Fernández (2002, pp. 563–93). It is interesting to note their notion of 'tutored generations', referring to the young scientists who were strongly supported and promoted in the context of the JAE. Thus, for example, 'the Physics Research Laboratory was created for Blas Cabrera and the Laboratory Mathematical Seminar was set up for Julio Rey Pastor in 1915' (González and Fernández, 2002, pp. 570–71).

[18]   In fact, the internal regulations approved by the JAE in 1907 were suspended until June of the same year, when they were finally approved by the conservative government.

controls over the awarding of study grants. The JAE was now only able to propose a list of preferred subjects (priority was given to those not taught in Spain), applications were to be submitted through the 'teaching institutions to which the candidates belonged' or in which they were studying and they had to be accompanied by 'a report from the university's Senate' (GM, 30, 30 January 1909, pp. 263–4). In this way, the JAE's proposal was counterbalanced by the report from the universities (both of which had to be presented together). Thus, the government attacked on two fronts, implementing new measures and submitting applications through the universities over which it had considerable influence.

All these conservative measures ensured political control over the JAE and, therefore, posed a boundary problem.[19] The boundary between science and politics, which had been established by the liberal government, was weakened. Conservative policy largely dissolved this boundary and limited the autonomy of the JAE. Indeed, the new ministry attempted to reduce the JAE to a mere advisory body rather than an executive one.

Liberal reactions to the new government measures were swift. A parliamentary debate raged for several years, in fact throughout the entire existence of the JAE. The debate started with the former minister Amalio Gimeno (now in opposition) challenging the new minister Rodríguez San Pedro over the measures taken.[20] The liberals' main argument was to highlight the importance of the JAE and they pointed out the need to maintain its autonomy and self-management on the basis of its scientific authority. In this sense, Gimeno made the following claim: 'Your Honour has still not understood the thinking behind the creation of the Board, nor why it was set up, what its purpose is, what it means nor what it will achieve' (DSS, 1907, 93, session 14-XI, p. 1851).

Regarding the obligation to submit three candidates to the Minister for a final decision, Gimeno stated that this would result in arbitrariness and ministerial caprice, with the Minister being able 'to place a friend who is third, in the second or even the first place'. What Gimeno was arguing was that 'if these decisions were left up to politicians, the objectivity and integrity that characterized scientific decisions would be lost', since such decisions would be

---

[19]   The effort to control the JAE was common to the various conservative governments and the Dictatorship of Primo de Rivera (1923–31). Under the dictatorship some services of the JAE were suspended, the number of grants was reduced and government control was increased. In 1926 a Royal Decree changed the system of designation of JAE government members (vocals) and the Minister got the power to nominate half of them to ensure greater control. The freedom of choice of its members by the JAE was recovered in 1930.

[20]   See this debate in Laporta, Ruíz, Zapatero and Solana (1987a, pp. 18–31).

subject to networks of influence (DSS, 1907, 93, session 14-XI, p. 1851). The scientists, whose integrity, objectivity and authority were argued to be far above that of the Minister, were the people who should decide to whom the grants should be awarded.

The liberals were convinced that scientists guaranteed the good government of science; on the contrary conservative politicians believed that politicians had much to say in the government of science. Moreover, the conservatives considered that the JAE was a liberal political project (the brainchild of Krausists from the ILE) and that decisions made in its context were political and ideological in nature, regardless of the fact that they were prestigious scientists.

These two opposing ways of understanding the JAE clashed repeatedly throughout the whole of the institution's existence.[21] The conservative governments were characterised by successive attempts (some more successful than others) to control the JAE, while the liberal governments attempted to restore the autonomy of the original project (although some slight problems of control also arose).[22]

The debate between liberals and conservatives served to clearly define the different ideas held by the two parties regarding the relationship between the State and science. For the liberals, science and the State were two separate spheres which collaborated with one another in order to achieve a public good necessary for the development and progress of the country, and this collaboration was based on mutual trust and implied autonomy. The internal values of science were considered far better suited than political ones to guide all aspects of science, including its governance. For the conservatives, this trust was unfounded, since the governance of science was mostly based on political criteria and in fact this is what happened in the JAE, due to the interests and liberal ideology dominant in the Board. Consequently: a) administration of resources and management

---

[21]   A period marked by the alternating of liberal and conservative governments until 1923, then by Primo de Rivera's dictatorship (1923–31) and finally by the Second Republic (1931) until the end of the Spanish Civil War in 1939. During the Second Republic (1931–39), the JAE operated unhampered. Indeed, it awakened much sympathy among republican politicians and the republican governments dealt with it generously and liberally. In fact, the Parliament of the Second Republic contained a high proportion of intellectuals and university lecturers, hence the name 'the Republic of the Intellectuals' or the 'Professors' Republic'. Among these, a significant number holding parliamentary seats were scientists, such as Negrín and Marañón.

[22]   There were exceptions to this general tendency: some conservatives were favourable to the JAE, such as Zacarías García, a Jesuit collaborator of the Centre for Historical Studies who praised the work of this Centre in the article 'El orden en los institutos y universidades del Estado', published in 1923 in the Journal Fe y Razón, 67 (López, 2006, p. 92).

of science could not be left entirely in the hands of scientists, politics must control and intervene; and b) the idea of an institution which was truly separate from partisan and even group interests was inconceivable, no matter how much JAE scientists claimed otherwise. The debate made it quite clear that, for conservatives, the idea of separating science and politics was not acceptable and was in itself a political idea.

Integrity Problems: the Partisanship Question

The question of partisanship was raised again and again by conservatives throughout the entire existence of the JAE, which was accused of operating on the basis of a network of political and personal relations which were both exclusive and partial. On this point, the conservatives seem to be right since, although the JAE claimed not to be partisan, this aim was only partially achieved. Firstly because the JAE could not avoid its political alignment with the liberals (it was a liberal project after all), and secondly, because in fact there were networks of relationships and influence, at least in the granting of scholarships for study abroad. This situation resulted from the autonomy granted to the JAE, coupled with the fact that it operated outside the reach of the university hierarchies and consolidated procedures, which meant that it was able, to a large extent, to act at its own discretion.

This ability to exercise discretion was made clear in the criteria established by the JAE for awarding study grants. These criteria included objective aspects, such as the presentation of the merits referred to in the application, but they also, fairly prominently, featured subjective gauges, consisting of qualitative, and therefore discretional, considerations. José Castillejo (the secretary of the JAE) referred to this point, stating that the awarding of study grants 'changed its nature [...] replacing momentary, purely intellectual examination, or the comparison of official merits, with a judgement in which these elements may be treated, at the very most, as mere indications, since the decision should be based on a collection of individual scientific and moral circumstances, considerations of aptitude and vocation, as well as objective social and pedagogic concerns' (JAE, 1908–1909, p. 16).

This network of relationships can be detected in the correspondence by Castillejo with various people in relation to important aspects of the JAE. Thus, for example, Castillejo wrote to Cossío, one of the most outstanding members of the ILE, in 1907 about the list which had been drawn up of 72 grant applicants:

'I would appreciate it if you could find the time to glance over it, because you will know many of them, especially the teachers, and could perhaps give me confidential reports about those whom you deem to be eligible and those you deem not. *I am fully convinced that applications and works presented, and so on, are worth nothing in comparison with ten minutes of conversation* [my emphasis]. One of those who had presented the most pamphlets, books and studies came to see me yesterday and left me frankly disenchanted. He was a freeloader of the worst kind, with neither ideals nor training. As you will know many of the applicants *de visu* or will have reliable references for them, I believe your opinion to be indispensable. Please rest assured that I will exercise the utmost care in the use I make of any information you may provide me with'. (Castillejo, 1997, p. 382)

Even during the period of conservative government intervention, this network clearly remained active with regard to both the foreign study grants and other aspects of the Board's organisation. One example of this can be found in an amicable letter written by Josep Pijoán (secretary of the Institute of Catalan Studies) to Castillejo (21 June 1907) requesting a grant while ostensibly discussing plans for the summer: 'What will you do? If Don Francisco [Giner de los Ríos] were to go to an acceptable location, I may drop by, if your Board were to award me a grant to go to Paris for four months' (Castillejo, 1997, p. 349).

Another interesting example is a letter sent by Elvira Alonso to Castillejo from Paris in response to a friendly letter from him: 'As I remarked earlier, I have a keen interest in going to England, and in this case would hope that the person you send to Toulouse will be completely trustworthy' (Castillejo, 1997, p. 365).

On 18 August 1907, Castillejo wrote to Giner about certain organisational matters pointing out some names: 'Dear Don Francisco: To which trustworthy person shall we turn for matters of music (!) Perhaps to Roda? Would he be a trustworthy person to be proposed for the shortlist of Board member candidates? I have not had sufficient contact with him. And to whom shall we turn for matters of architecture in Madrid? And, what about the arts and crafts, and industrial art?' (Castillejo, 1997, p. 368). A little later he goes on to say: 'To take youngsters, if they so wish, paying them travel and living expenses, and so on: [I propose] del Río, Barnés, Palacios, Zulueta, Acebal, and so on. Cajal said yes to everything' (Castillejo, 1997, p. 370).[23]

In the same letter, Castillejo made the question of recommendations explicit and wrote to Giner: 'I receive letters of recommendation (not many). I have

---

[23]   This makes it clear, as Sánchez Ron notes (2010, p. 106), that 'Castillejo knew how to deal with both Cajal and his superiors in the Ministry'.

decided to answer them' (Castillejo, 1997, p. 371). Giner warned him to be careful with recommendations in order not to offend anyone, and reminded him that Luis Montoto (an ecclesiastic notary and member of the Seville City Council) was recommended and so obtained his grant (Castillejo, 1997, p. 371).

In Montoto's case, it was his father who wrote a letter to Santiago Magdalena (Dean of the Saint Prioral Church in Ciudad Real, who knew Castillejo) asking him for a recommendation for a grant for his son (8 August 1907): 'Well, this is a stroke of luck! Mr Castillejo is the secretary and soul of the Board for Advanced Studies and Scientific Research which awards foreign study grants. I therefore have high hopes for your recommendation' (Castillejo, 1997, p. 371). Santiago Magdalena wrote in no uncertain terms to Castillejo regarding this recommendation: 'My distinguished friend, some days ago a colleague of mine from Seville asked me to recommend to you his son, who hopes to receive a grant to study or extend his studies abroad, and I thought you might be able to help our case (...)' (Castillejo, 1997, p. 372). He then outlined the young man's value and added: 'I would be most grateful if, upon my recommendation, Luis Montoto y Sedas could be preferred over other candidates, and in anticipation of that I extend my heartfelt gratitude and my warmest salutations to your family. Your affectionate friend (...)' (Castillejo, 1997, p. 372).

One of the networks of influence that the conservatives often referred to in their critiques was the web of relationships which linked the ILE and the JAE (what has been termed a 'diffuse institution').[24] This network was particularly active in the awarding of study grants, so much so, in fact, that the number of grants awarded to ILE members became alarmingly high, causing Acebal (Castillejo's assistant at the JAE secretariat) to sound the alarm. According to David Castillejo (José Castillejo's son): 'Acebal went on holiday and from Asturias sounded the first warning bell: there would be no one left at the Institution (ILE) if they all left to study abroad. This warning hid a deeper fear: that too many people at the Institution were taking advantage of the grants [...] Giner reacted energetically; it was, perhaps, the only occasion on which he felt morally criticized for lack of impartiality' (Castillejo, 1997, p. 377). Acebal wrote to Castillejo: 'I am a little (or perhaps more than a little) concerned over the number of lecturers at the Institution (ILE) who are travelling out of Spain; I foresee we shall have to act in order to reinforce the organization' (Castillejo, 1997, p. 378).

---

[24]   This term was introduced by Zulueta to refer to the influence of the ILE on other institutions through people educated in it (Zulueta, 1915).

A much later example, which alongside others demonstrates the continuity of this network of influence, was the grant awarded to Rafael Alberti, the well-known Spanish poet. In August 1931, Alberti, who was close because of familiar ties to the JAE member Menéndez Pidal, wrote to him from Paris requesting his aid for a grant (Sánchez Ron, 2010, p. 140). Pidal wrote a request to the Chairman of the JAE and the Board agreed to award Alberti a grant to spend four months in France. As Sánchez Ron points out 'the corresponding JAE file contains only a brief note with no date' (Sánchez Ron, 2010, p. 141). Subsequently, Alberti obtained a six-month extension of his grant to travel around Germany, Poland and Russia.

This preferential treatment received by those who knew the right people poses the question of the extent to which this set up hampered and thwarted the applications of those outside this network of personal relations. Correspondence between those involved reveals that Gimeno's fear that the conservative Minister could prioritise some grant applications over others for personal, non-objective reasons, extended to the JAE itself. From the perspective of the social contract for science, what was at stake here was a problem of 'integrity' in the management of resources which contradicted the principles and values that 'theoretically' embodied the JAE and that were assumed by their members. On the other hand, the problem of integrity extended to the management of public funds associated with these grants, that is, it had economic consequences.

Of course, the action of networks of this kind must be placed in its proper historical context and we should not forget that this was not an uncommon practice at the time. However, firstly, this does not eliminate the aforementioned problem of integrity and, secondly, the members of the JAE, in theory at least, rejected this practice and the values associated with it, publicly defending the characteristic values of the 'morality of science' which, as we have seen, from the end of the nineteenth century onwards, were the hallmark of the regeneration and modernisation of Spanish science, of which the JAE was the flagship.

It might also be possible to downplay the problem of integrity by claiming that it involved just a few members of the JAE. However, although the networks may have operated through a few select people, it is difficult to exonerate the institution since the proposals for study grants were drafted by the Executive Committee, and these proposals were approved by the Plenary Board, consisting 21 members and the secretary.[25] What this shows, therefore, is that neither

---

[25]   The members of the Executive Commitee included the Chairman of the JAE Santiago Ramón y Cajal, the secretary José Castillejo, the two Deputy Chairmen and two board members. According the Royal Decree of 22 June 1907, article 26: 'The Executive Committee shall review applications and make proposals [...] These proposals shall determine

institutions nor scientific communities act exclusively in accordance with the values assumed by the idealised concept of science that theoretically they sustain. This was the case of the JAE social contract for Spanish science.

Management of the Funds

Conservatives were critical for many years of the fact that the JAE's financial management was very liberal and, therefore, according to them, presented accountability problems. In 1912, the liberal minister Gimeno (then in the government) was accused by conservatives of squandering the general budget. In this context the Minister was criticised for the lack of budgetary control over the JAE and the use of credits for unspecified general purposes. Gimeno merely responded that the JAE had an allocation which it could administer freely, without this in any way breaking the law (DSC, 1911–12, 96, Session 29 February 1912, pp. 2633, 2626–7). However, what was really being attacked here was that the allocation was not subject to parliamentary control, and was granted with no strings attached for the Board to 'use it at its whim'.[26]

The issue arose once again in 1918 with a question posed by the conservative MP Pio Zabala (professor of history and future president of the University of Madrid under Francoism) in which he highlighted the JAE's success in including in the Budget Act 'an article which, with no specification, with no determination of the items administered by the Board, removes the precise destination of said sums from the Chambers' scrutiny using these general formulas: "For grant to Board of Pensions bound for general purposes 300,000 pesetas. For grants

---

for each study grant, the sum and period of the grant' (GN, 173, p. 1140); article 27 states that: 'The proposals so made, shall be submitted to the Plenary Board and with its approval, shall be elevated to the Ministry for a final decision regarding a list which shall include, whenever possible, three times as many names as grants'. This latter stipulation is the one the JAE argued so strongly against in relation to this Decree, which approved the JAE's internal regulations. In 1910, the now liberal government amended the previous decrees in a Royal Decree of 22 January 1910, which stated the following in relation to the aforementioned issue: Art. 26. 'The Executive Committee shall review applications and elevate to the Plenary Board [...] the draft proposals, specifying the sum and duration of each grant, and what shall be paid to cover travelling expenses'. Art. 27. 'Once approved by the Plenary Board, the draft proposals shall be sent to the Ministry for a final decision' (GM, 28, 28 January 1910, p. 199).

[26] 'This Board, Messrs deputies, has two kinds of allocations: one of 525,000 pesetas for pensions in and out of Spain, for studies abroad and other purposes set by the Royal Decree of creation [...] The Board could freely dispose of the other kind of allocation and yet it does not do, because they present justification of all expenses for the approval of the Minister' (DSC, 1911–12, 96, Session 29 February 1912, p. 2633).

payment and other expenses of the services, and so on, 334,000 pesetas".[27] Zabala also asked 'the Honourable Minister of Public Instruction to present the payrolls of those working on the Board and working at the Centres which depend upon it, as well as a specific list of the State salaries received by said people' (DSC, 1918, 22, Session 19 April 1918, p. 503). The integrist senator José María González de Echávarri presented a report from the National Audit Office with which he aimed to show that the JAE was not sufficiently accountable (Laporta, F., Ruíz, A., Zapatero, V. and Solana, J., 1987a, pp. 58–9). Similarly, the Senator and Professor of Sciences, Faustino Archilla stated that: 'The JAE enjoys a number of unjustified privileges, such as (among others) not having to submit its accounts to the National Audit Office as all other official centres must' (DSS, 1918, appendix 74, Session 12 July 1918, p. 1226).

The liberals' response was to deny all the accusations and claim that everything had been carried out in accordance with the law. The Senator José Casares responded to the criticism by acknowledging that the JAE had two types of allocations: one in the form of a subsidy which did not have to be justified and could be administered freely, and another which was subject to official justification. There were, therefore, certain sums which were free from any kind of control, although Casares maintained that he did not believe that this subsidy should be freely administered and stated that the Board did indeed present detailed records of how the money was used, even though it was not legally obliged to do so (DSS, 1918, appendix 6°, 74, Session 12 July 1918, p. 1219). The liberal Minister Santiago Alba, on the other hand (with whom Castillejo had an excellent relationship), reminded them of the JAE's independence from the Ministry of Public Instruction and the transparency of its accounts (DSS, 1918, appendix 6°, 74, Session 12 July 1918, p. 1224)

González de Echávarri maintained his initial position and pointed out that, from the responses provided, one could deduce that the JAE enjoyed certain privileges in the administration of its funds and that his proposed bill aimed to put an end to this situation (Laporta, F., Ruíz, A., Zapatero, V. and Solana, J., 1987a, p. 63). In fact, the question of how the JAE's economic resources were managed and the JAE's privileges in this field remained at the heart of parliamentary debates for years.

The accusations levelled by the conservatives were serious ones, and are indicative of the political and ideological tension which arose around the JAE.

---

[27]   Zabala affirmed: 'I was saying [...] that the Board of Pensions receives every year 700,000 to 800,000 pesetas, which can be administered with that wide margin of freedom that give it the decrees of its founding' (DSC, 1918, 22, Session 19 April 1918, p. 503).

However, it is true that the JAE was able to manage certain sums with freedom and was therefore in a privileged position with regard to the administration of its funds (despite being accountable). This had a very different meaning for conservatives than it did for liberals, due to their very different ways of understanding the project that the JAE embodied.

Finally, the JAE was also accused of unfair competition with universities, since it soaked up public funds for research and co-opted the best graduates, thus distancing research from the universities. These criticisms demonstrate the belief held by many that reform should have been carried out in the universities. The universities' principal problem was the perennial lack of funds which resulted in a lack of laboratories and libraries, and in turn hampered the pursuit of high-quality research. This state of affairs was openly denounced by many universities, including the University of Barcelona, one of the most progressive universities in the country, which also accused the JAE of centralism, since it confined its activities almost exclusively to Madrid (Laporta, F., Ruíz, A., Zapatero, V. and Solana, J., 1987a, p. 53ff.). The Spanish universities claimed that they could carry out high-quality research, as the JAE did, providing they had the necessary resources.

The disagreement, regarding whether or not it would have been possible to bring about the same degree of improvement of research that was obtained by the JAE if the universities had been reformed and had been the focus of investment policy, raged throughout the JAE's entire existence. Indeed, disagreement still exists today among scholars who have, as yet, been unable to come up with a definitive answer to this question.

Some Concluding Remarks

The debate between conservatives and liberals has been analysed by historians, and other experts, who generally have tended to consider conservative accusations as political and ideological attacks against the JAE. Nevertheless, as we have seen, this was not exactly the case in relation to the network of influence, the freedom with which the JAE was able to administer certain budgets, and the exclusion of the universities from the political project which aimed to modernise scientific research and education.

On the other hand, it should be noted that the integrity and boundary problems analysed were partial since scientific autonomy was called into question at certain stages of the history of the JAE, and integrity problems were not related to research but to scientific training and the resources required for it. Beyond the problems outlined above, the relationships between science and politics that determined Spanish scientific development in the first third of the twentieth century constituted a social contract for science which was productive. It was a major boost to Spanish scientific research, which began to generate original work that was recognised at an international level. Scientists had fulfilled their side of the contract regarding productivity, as evinced by the JAE's long list of achievements.

As we have seen, one of the JAE's basic aims was to provide grants to study abroad and indeed it dedicated a large part of its resources and efforts to achieving this goal.[28] During the 31 years for which the institution existed, it provided 1,723 grants to study abroad at different prestigious international universities and laboratories. In addition to providing high-quality scientific training, these grants also helped establish networks of scientific relations with prestigious centres in Europe – something of vital importance to Spanish scientists.

Another major objective was to set up well-equipped science centres and laboratories, in which to conduct high-quality basic research. The resulting facilities were pure research centres, staffed by grant holders and leading scientists. The following general data demonstrate the achievements of the JAE in this field. In 1910 the National Institute for Physical-Natural Science was established, chaired by Ramón y Cajal. That same year also was the foundation of both the Guadarrama Alpine Station and the Centre for Historical Studies chaired by Ramón Menéndez Pidal. The Students Residence (for men) was created in order to redress a historical lack of facilities for students from outside Madrid. The Residence was equipped with a number of different laboratories, the first of which was the Physics Research Laboratory directed by the great physicist Blas Cabrera. In 1912 the Palaeontology and Prehistory Research Commission was set up, and the Students Residence gained several new laboratories: the Microscope Laboratory, the General Chemistry Laboratory and the Nervous System Histopathology Laboratory. In 1915 the women's section of the Residence was established under the leadership of María de Maeztu; in this year the Physiological Chemistry and Mathematical Seminary Laboratories were also established. In 1916 the General Physiology Laboratory was set up and the Nerve Centre Physiology and Anatomy Laboratory was established in the male Residence. In 1918 the experimental High-School was created to train teachers in innovative teaching methods. Then in 1921 the Biology Mission in Galicia, in 1922 the Maritime Station in Pontevedra and in 1932 the National

---

28   Influence networks acted in this sphere, as noted earlier, but, of course, did not affect all grants.

Physics and Chemistry Institute were opened. This latter Institute was funded by the Rockefeller Foundation and directed by Blas Cabrera.

The JAE also implemented an interesting publications policy.[29] International conferences were held in Madrid, and eminent scientists travelled to Spain (including Marie Curie and Albert Einstein). Scientists such as Enrique Moles, Miguel Catalán, Blas Cabrera, Severo Ochoa and Antonio de Zulueta made original contributions in their fields, which were internationally recognised. Therefore, one of the JAE's most important achievements was its notable contribution to consolidating an internationally acknowledged Spanish scientific elite and which raised Spanish scientific research to hitherto unattainable heights.

To summarise, by the outbreak of the Spanish Civil War in 1936, Spanish science had reached a significant level of development and had started to close the gap with European science. For the first time in two centuries Spain was finally on the road to modernisation and scientific standardisation. Much still remained to be done and it was important to continue supporting scientific and technological development. But the social contract for Spanish science had borne valuable fruits and had a bright future.

All this, however, was cut short by the outbreak of the Spanish Civil War, which was won by precisely the most recalcitrant Catholic sectors that were virulently opposed to the modernisation programme for Spain and Spanish science and thought.[30] Indeed, they believed that this programme was the worst problem affecting Spain and was responsible for denaturing and breaking up the country. They considered modern science and technology, along with the positive and rationalist mentality and the Europeanisation of ideas, to be profoundly wrong and completely alien to what they saw as the true Spain. Their alternative was to return to imperial, ultra-Catholic values and to subordinate science to faith. This not only broke the social contract for Spanish science, it also eliminated for decades the whole modernisation and Europeanisation process for Spanish science, along with the effort to achieve high-quality scientific and technical research (Gómez and Canales, 2009).

---

[29] Research papers by scholars were published in the *Annals* (XIX vols). The JAE also published Journals such as: *Revista de Filología Española* (founded by Menéndez Pidal in 1914), *Archivos de Neurobiología* (founded by Lafora, Ortega y Gasset and Sacristán in 1920), *Archivo Español de Arte y Arqueología* (founded by Manuel Gómez Moreno in 1925), *Residencia* (1926–35). Books were also published, including: *Fauna Ibérica: Mamíferos* (Ángel Cabrera) 1914; *Meditaciones del Quijote* (Ortega y Gasset) 1914; *La Herencia Mendeliana* (José Nonídez) 1922; *Principio de Relatividad* (Blas Cabrera) 1923 and *Orígenes del Español* (Menéndez Pidal) 1926.

[30] As Canales (2012) points out, opposition was so strong that Francoism even sought to erase any physical trace of the JAE, building a new campus on its premises.

List of References

Álvarez, P.F. and Vázquez-Romero, J.M., 2005. *Krause, Giner y la Institución Libre de Enseñanza. Nuevos estudios*. Madrid: Universidad Pontificia Comillas.

Balmer, B., 2001. *Britain and Biological Warfare: Expert Advice and Science Policy, 1930–65*. Basingstoke: Palgrave.

Balmer, B., 2012. *Secrecy and Science: A Historical Sociology of Biological and Chemical Warfare*. Farnham: Ashgate.

Bayles, M., 1983. *Professional ethics*. Wadsworth: Belmont.

Bush, V., 1945. *Science – The Endless Frontier: A Report to the President on a Program for Postwar Scientific Research*. Washington: US Government Printing Office.

Cacho Viu, V., 1962. *La Institución libre de enseñanza. I Orígenes y etapa universitaria (1860–1881)*. Madrid: Ediciones Rialp.

Canales Serrano, A.F., 2012. 'A new space for a new science: the transformation of the JAE Campus after the Spanish civil war'. *History of Education*, 41(5), pp. 657–74.

Canales Serrano, A.F., 2013. 'La modernización del sistema educativo español, 1898–1936'. *Bordón*, 65(4), pp. 105–18.

Caso J.M. (ed.), 1989. *El Censor*. Oviedo: Instituto Feijoo de Estudios del Siglo XVIII.

Castillejo, D. (ed.), 1997. *Un puente hacia Europa, 1896–1909. Epistolario de José Castillejo. I*. Madrid: Castalia.

Castillejo, J., 1976. *Guerra de ideas en España*. Madrid: Biblioteca de la Revista de Occidente.

Cavanilles, A.J., 1784. 'Observaciones del abate Cavanilles sobre el artículo "España" de la Nueva Enciclopedia'. In: E. García Camarero and E. García Camarero (eds), 1970. *La polémica de la ciencia española*. Madrid: Alianza, pp. 54–7.

Crowther, J.G., 1949. *Science in liberated Europe*. London: The Pilot Press.

Crowther, J.G., Howarth, O.J.R. and Riley, D.P., 1942. *Science and world order*. England, US: Penguin Books.

Demeritt, D., 2000. 'The New Social Contract for Science: Accountability, Relevance, and Value in US and UK Science and Research Policy'. *Antipode*, 32(3), pp. 308–29.

Denina, C., 1786. *Respuesta a la pregunta ¿Qué se debe a España? (Discurso leído en la Academia de Berlín, en la Asamblea pública del 26 de enero, del año 1876, en el día del aniversario del Rey por el Sr. Abate Denina)*. Madrid: Imprenta Real.

Díaz, E., 1973. *La filosofía social del Krausismo español*. Madrid: Edicusa.

Echegaray, J., 1866. *Historia de las matemáticas puras en nuestra España (Discurso leído en la Real Academia de Ciencias)*. Madrid: Aguado.

Edgerton, D., 2006. *Warfare State: Britain 1920–1970*. Cambridge: Cambridge University Press.

Gener, P., 1877. *Herejías, Estudios de crítica inductiva sobre asuntos españoles*. Madrid: Ed. Fernando Fe.

Gener, P., 1903. *Cosas de España: Herejías nacionales. El renacimiento de Cataluña*. Barcelona: Juan Llordachs Librero.

Gibbons, M., 1999. 'Science's New Social Contract with Society'. *Nature*, 402(6761), pp. 81–94.

Gil, C., 1975. *Krausistas y liberales*. Madrid: Seminarios y Ediciones.

Giner, H., 1912. *Prólogo a tesis de Tiberghein*. Valencia: Sempere.

Glick, Th. F., 1982. *Darwin en España*. Barcelona: Ediciones Peninsula.

Gómez, A., 2014. 'Frontera e integridad en el contrato social para la ciencia española, 1907–1939'. *Dymanis*, 34(2), pp. 465–87.

Gómez, A. and Balmer, B., 2013. 'Ciencia y política: una cuestión de frontera'. In: A. Gómez and A.F. Canales (eds), 2013. *Estudios Políticos de la ciencia. Política y Desarrollo científico en el siglo XX*. Madrid: Plaza y Valdés, pp. 15–34.

Gómez, A. and Canales, A.F., 2009. 'The rebels and the new Spanish scientific culture'. *Journal of War and Culture Studies*, 2(3), pp. 321–33.

González, F. and Fernández, R.A., 2002. 'Nuevas perspectivas en torno a la política de pensiones de la Junta para Ampliación de Estudios: modelos de encuentro con Europa de la universidad española'. *Revista Complutense de Educación*, 13(2), pp. 563–93.

Guston, H.D., 2000. *Between Politics and Science. Assuring the Integrity and Productivity of Research*. Cambridge: Cambridge University Press.

Guston, H.D. and Keniston, K. (eds), 1994. *The Fragile Contract. University Science and Federal Government*. Massachusetts: The MIT Press.

JAE, 1908–1909. *Memoria correspondiente al año 1907*. Madrid: Hijos de M. Tello.

Jiménez, A., 1985. *El Krausismo y la institución libre de enseñanza*. Madrid: Cincel.

Kleinman, D.L., 1995. *Politics on the Endless Frontier. Postwar Research Policy in the United States*. Durham and London: Duke University Press.

Laporta, F., Ruíz, A., Zapatero, V. and Solana, J., 1987. 'Los orígenes culturales de la Junta para Ampliación de Estudios'. *Arbor*, 126 (493), pp. 17–87.

Laporta, F., Ruíz, A., Zapatero, V. and Solana, J., 1987a. 'Los orígenes culturales de la Junta para Ampliación de Estudios'. *Arbor*, 127 (499), pp. 9–137.

López, Sánchez, J.M., 2006. *Heterodoxos españoles: el Centro de Estudios Históricos, 1910–1936*. Madrid: Marcial Pons.

Lubchenco, J., 1998. 'Entering the Century of the Environment: A New Social Contract for Science'. *Science*, 279(5350), pp. 491–7.

Marañón, G., 1941. 'Nuestro siglo XVIII y las academias'. In: E. García Camarero and E. García Camarero (eds), 1970. *La polémica de la ciencia española*. Madrid: Alianza, pp. 487–516.

Masson de Morvilliers, N., 1782. 'España'. In: E. García Camarero and E. García Camarero (eds), 1970. *La polémica de la ciencia española*. Madrid: Alianza, pp. 47–53.

Menéndez y Pelayo, M., 1894. 'Esplendor y decadencia de la cultura científica española'. In: E. García Camarero and E. García Camarero (eds), 1970. *La polémica de la ciencia española*. Madrid: Alianza, pp. 311–50.

Merton R., 1974. *The Sociology of Science: Theoretical and Empirical Investigations*. Chicago: University of Chicago Press.

Mukerji, C., 1989. *The Fragile Contract: Scientists and the State*, Princeton: Princeton University Press.

Mulkay, M., 1973. 'Norms and ideology in science'. *Social Science Information*, 4(5), pp. 637–56.

Núñez, R.D., 1975. *La mentalidad positiva en España: desarrollo y crisis*. Madrid: Tucar Ediciones.

Núñez, R.D., 1977. *El darwinismo en España*. Madrid: Castalia.

Otero Carvajal, L.E. and López, J.M., 2012. *La lucha por la modernidad. Las ciencias naturales y la Junta para la Ampliación de Estudios. Madrid: Consejo Superior de Investigaciones Científicas*. Madrid: Publicaciones de la Residencia de Estudiantes.

Picatoste, F., 1866. 'El discurso del señor Echegaray en la Academia de las Ciencias'. In: E. García Camarero and E. García Camarero (eds), 1970. *La polémica de la ciencia española*. Madrid: Alianza, pp. 191–7.

Pielke, J.R., 2007. *The Honest Broker. Making Sense of Science in Policiy and Politics*. Cambridge: Cambridge University Press.

Polanyi, M., 1962. 'The Republic of Science Its Political and Economic Theory'. *Minerva*, 1, pp. 54–73.

Price, D.K., 1954. *Government and Science: Their Dynamic Relation in American Democracy*. New York: New York University Press.

Ramón y Cajal, S., 1897. 'Deberes del Estado en relación con la producción científica (Discurso de ingreso en la Real Academia de Ciencias)'. In: E. García Camarero and E. García Camarero (eds), 1970. *La polémica de la ciencia española*. Madrid: Alianza, pp. 373–99.

Ramón y Cajal, S., 1923. *Historia de mi Labor científica. Neuronismo o reticularismo*. Barcelona: Biblioteca Universal del Cículo de Lectores, 1997.

Ramón y Cajal, S., 1941. *Reglas y Consejos sobre Investigación científica. Los tónicos de la voluntad*. Madrid: Austral, 2007.

Revilla, J., 1854. *Breve reseña del estado presente de la Instrucción Pública en España, con relación especial a los estudios de filosofía*. Madrid: Imp. Por Eugenio Aguado.

Rey Pastor, J., 1915. 'El progreso de España en las Ciencias y el progreso de las Ciencias en España'. In: E. García Camarero and E. García Camarero (eds), 1970. *La polémica de la ciencia española*. Madrid: Alianza, pp. 458–78.

Rodríguez Carracido, J., 1897. 'Condiciones de España para el cultivo de las ciencias'. In: J. Carracido (ed.), 1987. *Estudios Histórico-Críticos de la Ciencia Española*. Madrid: Imprenta de Alrededor del Mundo, pp. 19–47.

Rodríguez Carracido, J., 1911. *El problema de la investigación científica en España*. Madrid: Imprenta de Eduardo Arias.

Rodríguez Carracido, J., 1911a. 'La enseñanza de la química en España'. In: A. Moreno and A. Josa (comps), 1988. *José Rodríguez Carracido. Estudios Histórico-Críticos de la Ciencia Española*. Barcelona: Ed. Alta Fulla, pp. 397–403.

Rodríguez Carracido, J., 1911b. 'Cómo cultivamos la Química en España y cómo debe ser cultivada'. In: A. Moreno and A. Josa (comps), 1988. *José Rodríguez Carracido. Estudios Histórico-Críticos de la Ciencia Española*. Barcelona: Ed. Alta Fulla, pp. 397–403.

Romero de Pablos, A. and Santesmases, M.J. (eds), 2008. *Cien años de política científica en España*. Madrid: Fundación BBVA.

Sánchez Ron, J.M. (ed.), 1988. *La Junta para Ampliación de Estudios e Investigaciones Científicas 80 años después, 1907–1987*. Madrid: CSIC.

Sánchez Ron J.M., 1998. *Un siglo de ciencia en España*. Madrid: Publicaciones de la Residencia de Estudiantes.

Sánchez Ron, J.M., 1999. *Cincel. Martillo y piedra. Historia de las ciencias en España (siglos XIX–XX)*. Madrid: Taurus.

Sánchez Ron, J.M., 2010. 'Encuentros y desencuentros: relaciones personales en la JAE'. In: Sánchez Ron, J.M. and García-Velasco, J. (eds), 2010. *100 años de la JAE. La Junta para Ampliación de Estudios e Investigaciones Científicas en su Centenario*. Madrid: Publicaciones de la Residencia de Estudiantes – Fundación Francisco Giner de los Ríos – ILE, pp. 94–215.

Sánchez Ron, J.M., Lafuente, A., Romero, A. and Sánchez de Andrés, L. (eds), 2007. *El Laboratorio de España. La Junta para Ampliación de Estudios e Investigaciones Científicas 1907–1939*. Madrid: Sociedad Estatal de Conmemoraciones Culturales/Amigos de la Residencia de Estudiantes.

Santesmases, M.J., 2008. 'Orígenes de la política científica en Europa'. In: Romero de Pablos, A. and Santesmases, M.J. (eds), 2008. *Cien años de política científica en España*. Madrid: Fundación BBVA, pp. 293–322.

Sanz del Río, J., 1860. *Sistema de la filosofía. Metafísica. Primera parte: análisis*. Madrid: Manuel Galiano.

Sarewitz, D., 1996. *Frontiers of illusion: Science, technology and the politics of progress*. Philadelphia: Temple University Press.

Slaughter, S. and Rhoades, G., 2005. 'From "Endless Frontier" to "Basic Science for Use": Social Contracts between Science and Society'. *Science, Technology, & Human Values*, 30 (4), pp. 536–72.

Steelman, J.R., 1947. *Science and Public Policy*. Washington, DC: U.S.G.P.O.

Tiberghien, G., 1875. *Introducción a la Filosofía y preparación a la Metafísica*. Madrid: J. Morales.

Zulueta, L., 1915. 'Lo que nos deja D. Francisco'. *Boletín de la Institución Libre de Enseñanza*, 39, pp. 48–56.

Serial References

DSC: *Diario de Sesiones del Congreso de los Diputados*
DSS: *Diario de Sesiones del Senado*
GM: *Gaceta de Madrid*

Chapter 3

Spanish Science: from the Convergence with Europe to Purge and Exile[1]

Francisco A. González Redondo

From Civil War to Reconciliation: a Preliminary Overview

In 1976, Democracy was reinstated in Spain. The gates for peace and reconciliation seemed to be opened for Spanish science after forty years of personal and social tragedy caused by civil war, purge and exile. Regrettably, many scientists had died by then, and, in fact, only two exiled science professors were young enough in 1936 so as not to have reached the age of retirement in 1976: Francisco Giral González and Augusto Pérez-Vitoria. Thus, they could recoup their chairs and shared a remarkable welcome from those colleagues that stayed in their country and had made a career in Francoist Spain (Giral, 1994).

It is true that several initiatives had already allowed high profile physicists to return and be partially reconciled, for example Arturo Duperier (exiled in London) during the early 1950s and Manuel Tagüeña (exiled in Mexico) at the beginning of the 1960s. In addition, leading mathematicians Julio Rey Pastor and Esteban Terradas were allowed to return (from Argentina) at the end of 1939. On the other hand, several scientists such as chemist Enrique Moles attended the Regime's call to 'all Spanish citizens of good will living abroad', only to be imprisoned and judged after he crossed the border from France in 1941 (Berrojo, 1980).

Indeed, Spanish scientists manifested very different attitudes towards the two sides of the conflict between July 1936 and March 1939. This was sometimes because of their personal convictions, for others it was a result of their location in Spain after the military plot. And, the consequences of their particular compromises and/or disaffections would also differ (González Redondo, 2013). Some scientists remained loyal to the Republic, as did physicist Pedro Carrasco;

[1]   This article was made possible thanks to the Spanish National Research Project FFI2012–33998 (Ministry of Economy and Competitiveness).

others joined Franco's newly organised institutions, as did paediatrician Enrique Suñer. Some tried to follow a discrete life in order to survive in nationalist Spain, waiting for an impossible republican victory, for instance the physicist Miguel Catalán; while others adapted themselves to Republican Spain while waiting for the predictable nationalist win, as did chemist Ángel del Campo. Some scientists, such as physicist Julio Palacios, even worked as Franco's fifth-columnists in besieged Madrid. Finally, a group of significant personalities, including philosopher José Ortega y Gasset, pedagogue Luis de Zulueta, physicist Blas Cabrera and historian Claudio Sánchez Albornoz, decided to leave their country just as the war began, constituting what became known as the *Third Spain* (González Redondo and Fernández Terán, 2010).

This heterogeneity of responses to the war was a natural outgrowth of the different attitudes towards internal regeneration and external convergence with Europe developed during the first decades of the twentieth century. These were the years running from the *1898 disaster*, when Spain lost the remains of its former colonial empire, to its encounter with the most scientifically advanced of its neighbouring nations after the advent of the Second Republic and until 1936.

This journey towards a modern Spain, a country suffering several decades of underdevelopment when compared to the rest of Europe, had two personal models that could serve as guiding references for the rest of Spanish scientists. On the one hand, Santiago Ramón y Cajal, who was awarded the Nobel Prize for Physiology in 1906, proved that bio-medical sciences could keep up with the times. On the other, Engineer Leonardo Torres Quevedo's pioneering contribution, in the turn from the nineteenth to the twentieth century, had resulted in him being characterised as 'the most prodigious inventor of his time' (d'Ocagne, 1930), a great honour for applied Spanish sciences and technology.

Together with personal convergence, *regenerationist* governments understood that institutional convergence was also needed. Thus, the Institute for Biological Studies (*Instituto de Investigaciones Biológicas*) was created for Cajal in 1901, while Torres Quevedo could enjoy the directorship of the Centre for Aeronautical Research (*Centro de Ensayos de Aeronáutica*) from 1904. Overcoming the nineteenth century's individualist isolation, science could be practised in Spain, now following European standards: scientific communities working in scientific institutions (González Redondo, 2004).

Nevertheless, there were attempts to create a positive intellectual atmosphere, the *spirit* of which emanated from the Free Institution for Teaching (*Institución Libre de Enseñanza* – ILE) (Jiménez-Landi, 1996). The origin of the *Institution* was based upon a radical intellectual crisis, felt not only nor first in Spain, but common to all of the Western World: the reception of Charles Darwin's *On the Origin of Species* (1859).

It was in the context of the international reaction against such 'dangerous' scientific beliefs that threatened well established western religious and cultural foundations that the ILE was created in 1876 by a group of Spanish professors who had been expelled from their chairs by the conservative government for explaining Darwin's ideas to their pupils (Pratt, 2001; Glick, 2001). In fact, and although initially conceived as a free autonomous college independent from official State-ruled universities, the *Institution* would only materialise as a private primary and secondary school once those professors were allowed to return to their positions. Its original aims and scope (its *spirit*) evolved so as to be identified with the essential and general strategy needed in Spain for its scientific and educational progress towards convergence with Europe, especially after the *1898 disaster*.

As might be expected, since its creation the *Institution* became an enemy for those defending tradition, and this was felt especially after the proclamation of the Second Republic and continuing through to the Civil War. Indeed, from 1936 the ILE was blamed for the tragedy suffered in Spain, so that the goal of removing all of its achievements and influence was taken as the ideological basis for the purge suffered by all those who compromised with the modernisation of their country, whether or not those intellectuals had been related to the *Institution*. The purge threw a significant part of Spain's scientific community into such a long lasting exile that it was impossible for most of them to witness the end of the Dictatorship, the advent of Democracy, and the beginning of that process of reconciliation that Giral and Pérez-Vitoria experienced from 1976.

Purge on the Eve of the Civil War: the Appeal to *Tradition*

On 8 November 1936, not yet four months after the military coup had become a complete civil war, Francisco Franco, the *Generalísimo* of the Nationalist Army, signed and sent for publication in the Official Gazette [*Boletín Oficial del Estado*] the provision which constituted the starting point for the purge of Spanish academics at all levels:

> The fact that for several decades academia, in all its stages and with increasingly rare exceptions, have been influenced and almost monopolized by dissolving ideologies and institutions, in open opposition to the national wit and tradition, makes it necessary that, in these solemn moments we are experiencing, a complete

and thorough revision of the personnel of Public Instruction should be undertaken, a formality that must be prior to a radical and definitive reorganization of all teachings, removing from its roots through this process those false doctrines that, together with its apostles, have been the main factors contributing to the tragic situation the Nation has been led into (BOE, 11 November 1936).

Franco was by no means capable of writing a paragraph like this by himself. This radical synthesis of such a well established line of thought characteristic of the rebels had been partially stated during the early 1930s by some ideologues who believed that 'the main factors for the tragic situation' of our Nation had been those 'dissolving ideologies and institutions'. Such a perspective can be found in the work of all those associated with the Marqués del Quintanar, Ramiro de Maeztu and Eugenio Vegas Latapie, in the journal *Acción Española* [Spanish Action] from which the Catholic, reactionary and anti-Republican party *Acción Nacional* [National Action] emerged in 1931 (Morodo, 1985).

In fact, the purge undertaken by Franco's Regime would be a special case in the context of the reactionary Europe of that time. So, prosecution of Jewish academics in Spain would be more a rhetorical element than an effective process. Furthermore, the purge and especially the General Cause conducted against those who lost the Spanish Civil War, would become models for the trials held by the allied forces after the Second World War against such defeated countries as Fascist Italy and Hungary, Nazi Germany and Imperialist Japan.

In fact, César Silió, former Minister of Public Instruction during the conservative Antonio Maura's and José Sánchez Guerra's Presidencies, had also advanced a particular Spanish nationalist perspective in 1933. For him, the summit of our 'national wit and tradition' was placed in the Spanish Golden Age, that is, the sixteenth century and first two thirds of the seventeenth century, an inheritance that our Europeanised intellectuals would deeply loathe:

> Spain had lost its colonies. It had surrendered almost without a fight. Those intellectuals sermonize that we should disown the golden legend and Europeanize ourselves, while spreading pessimism across Spain. The Spanish, now that they have lost their faith and patriotism and are resigned to their fate (Silió, 1933, p. 59).

But more precisely, it would be the ILE which was to be blamed as essentially the cause of the Civil War. In fact, this attribution rooted in works such as *Una poderosa fuerza secreta. La Institución Libre de Enseñanza* [A powerful secret power. The Free Institution for Teaching]. This book, which collected together several articles published during the Civil War, had as its primary intention to *denounce* the routes of infiltration and propagation followed by the ILE's *pernicious* ideas that had been *rarefying* in academia.

To achieve this purpose, two quotes were introduced in its Preface as supposed self-confessions of the offences committed by the *institutionists*. These two *absolute truths* were advanced in order to impute to them the crime of having enjoyed the power to govern Spanish life through an elite able to control the will of the people. They would even allow a spurious identification between liberal-reformist *institutionists* and the so much hated socialists. The first quote, with the editor's own commentary, read as follows:

> 'To the Red Revolution Socialism has given the masses while the *Free Institution for Teaching* has provided the leaders'. A great truth impressed by *El Socialista* in Madrid, in the boastful days of the republican victory. A truth subscribed by any man who, gifted with real historical sense, has observed what has happened in Spain. A truth confirmed by facts and proclaimed by the same conspicuous *institutionists* (Martín-Sánchez, 1940, p. 7).

The second quote in the same Preface, with new comments by the editor, transcribed a speech delivered by the republican Minister of Public Instruction, socialist Fernando de los Ríos, devoted to praise the work of the *Institution*:

> So the euphoria of power released the otherwise cautious tongue of Fernando de los Ríos, who went on to say: 'The expectancies of Giner de los Ríos's disciples were silently grafted onto the Spanish pedagogical organization. The Higher Teachers Training College [*Escuela Superior del Magisterio*], the Board for Advanced Studies and Scientific Research [*Junta para Ampliación de Estudios e Investigaciones Científicas* – JAE], the School of Criminology [*Escuela de Criminología*] and even the Students Residence [*Residencia de Estudiantes*] have become the seeds of the New Spain. These have been the germs that have made possible the advent of a new Regime. The seeds were lying quietly in the furrow. The Spanish Republic is harvesting their outgrowth' (Martín-Sánchez 1940, p. 7).

Having introduced a sample of those 'germs' in the Preface, the book would clarify, chapter by chapter, other 'dissolving institutions' where a purge should be immediately undertaken: the Experimental High School [*Instituto-Escuela*], the special State-supported College conceived as a test model for introducing the pedagogical principles of the ILE in Spanish High Schools; the Centre for Historical Studies [*Centro de Estudios Históricos*]; the National Institute of Physics and Chemistry [*Instituto Nacional de Física y Química*] and so on.

New chapters were also devoted to denouncing the influence of the ILE and its 'dissolving ideology' in such matters as teacher training and the provision of university chairs, their hoarding and spoiling of scientific research, and their support and control of summer universities.

Nevertheless, it really was in all those institutions where convergence with Europe was achieved, in a process which, indeed, would have removed the country from the path of that 'national wit and tradition' which began in the sixteenth century under the first Habsburg kings, a golden era for Spanish imperial expansion in America and Asia, and, of course, a flourishing period for Spanish Literature and the Arts. But this was also the century of the Protestant Reformation, and, in its second half, the century of the Council of Trent and its embodied Counter-Reformation, of which Spain and the Inquisition became patrons and custodians. Consequently, the ideological renewal intended by Erasmus of Rotterdam's Humanism for the Catholic kingdoms was dramatically cut during the reign of Philip II, whose religious fervour conditioned the closing of borders to any foreign influence that could endanger the orthodoxy set in Trent and the rules that emanated from Rome.

And it is in this framework that the appeal to *tradition* can be understood, as Miguel de Unamuno attempted in 1895 when he identified it with 'castism', characterised as 'the retaliation of the old national historic spirit reacting against Europeanization' (Unamuno, 1895, p. 39). For him, it constituted 'the work of the latent Inquisition' as 'the Inquisition was an instrument for isolation, for *castist* protectionism, for exclusionary individualization of the caste', which 'impeded here the flowering sprout of the reformed countries'. And he would conclude with his well known sentence: 'Spain still has to be discovered, and it will be discovered only by Europeanized Spanish' (Unamuno, 1895, p. 41).

Following this line of thought, *regenerationist* Joaquín Costa would remark that in the sixteenth century the European nations divided themselves in two groups: 'on one side, the future, the Modern Age of the World, represented by England, Italy, Germany, France; on the other, the past, a stubborn resistance to progress and new life, represented by Spain' (Costa, 1906, p. xvi).

Indeed, as Celestino Mutis wrote in 1762, Spain 'had remained arrested', far from the seventeenth century's Scientific Revolution (González de Posada, 2009), rejecting the new understanding of the World developed from Copernicus' seminal reorganisation of the Solar system, through Galileo's scientific apostolate, to Newton's formalisation of the mechanisation of the Universe. In fact, the peculiar reception and adaptation of the Enlightenment during the eighteenth century would not introduce much improvement to Spanish thought, as Jorge Juan (1774) would regret when considering the blindness of a nation that, 'opposing the Holy

Scriptures to the finest demonstrations in Geometry and Mechanics', had become 'the intellectual laughingstock of Europe' (Juan, 1774, p. 13).

The assumption of Newton's *System of the World* and its implicit explanation of the real place of the Earth in the Universe forced a reinterpretation of the Bible. This was something that could still not be assumed in those areas of the Spanish Empire such as actual Colombia where, as late as 1801, Celestino Mutis was prosecuted by the Inquisition for explaining Copernican heliocentric views which had been accepted and even been taught in Rome for several decades (González de Posada, 2009).

Nevertheless, after 1859 the Churches and their reactionary and conservative companions met a much more radical problem. Charles Darwin's *On the Origin of Species* and his later *The Descent of Man* (1871) had brought the possibility for a complete rupture with our anachronistic past. As a matter of fact, if *homo sapiens* had emerged (evolved) from a previous human species, Adam and Eve would have never existed, and so they could have not committed the original sin. If Humankind was not carrying the original sin, it was not necessary for God the Father to send his Son to the Earth to suffer martyrdom, die on the Cross and redeem us from original sin. In sum, although the existence of God was not explicitly put in doubt, after the last third of the nineteenth century the figure of Christ could be considered, implicitly, from a Darwinian perspective, as unnecessary. And indeed, in Spain it would be the ILE and those sharing its *spirit* who would undertake the task of modernisation of the nation through its scientific Europeanisation. This would certainly bring it into conflict with *tradition*.

On the Process of Scientific Convergence with Europe

Although the decline of the Spanish Empire could be foreseen since the reign of the last Habsburg kings and the arrival of the Bourbon dynasty, it became most evident after the Spanish war of independence from Napoleonic France and the subsequent Spanish American wars of independence during the first third of the nineteenth century. In fact, those timid efforts of regeneration undertaken in Spain during the last decades of the nineteenth century met an unsuspected ally: the *1898 disaster*. This was the loss of the last colonies in America (Cuba, Puerto Rico) and in Asia (the Philippines) to the USA. In short, an emerging power based upon science and technology had defeated a nation that had been for centuries a world reference, but in literature and the arts.

It is not surprising then that, when the liberal government in power at the time of the *disaster* resigned, one of the first resolutions adopted by the conservative party was the creation of the Ministry of Public Instruction (until 1900 just a department in the Ministry of Public Works). It at once assumed that education was to become the keystone for regeneration.

Taking as a starting point the laws and decrees enacted in just six months by this first conservative Minister, Antonio García Alix, the entire Spanish society engaged in the regenerative process. This was especially so when it was directly or indirectly inspired by the ideas of secularism, pedagogical innovation and Europeanisation advocated by the ILE. But, as argued in the previous chapter, all this was something that could only be achieved in the first decades of the twentieth century by liberals, reformists and republicans. It happened, for instance, in 1907 with the creation of the Board for Advanced Studies and Scientific Research (*Junta para Ampliación de Estudios e Investigaciones Científicas* – JAE). This was a public institution dependant on the Ministry of Public Instruction which awarded scholarships at leading research institutes in Europe to successive generations of talented Spanish students (GM, 15 January 1907, pp. 165–7). The Board, from 1910, received and integrated researchers into the two newly created institutions that led the way towards the desired convergence: the Centre for Historical Research (GM, 19 March 1910, pp. 582–3) and the National Institute of Physical and Natural Sciences (GM, 29 May 1910, pp. 410–11).

The *Centre* would become the *most Spanish* department of the JAE. In it, once European (especially German) historiographical theories, methodologies and techniques had been imported, learned, adapted and developed, its researchers could undertake a daunting task: to determine the *essence of Spain*. In other words, to answer the question 'What is Spain?' through a scientific approach to its history, art, language, philosophy and law (López Sánchez, 2006). Obviously, in those *regenerationist* years in which Spain, after the *1898 disaster*, was almost confined to its mainland limits, the *Centre* was called on to play a role in the process of the construction of a new, modern and European national identity. This was in direct conflict with the guardians of *tradition* and the old glories of Spain's imperial past. Tragically, these radical misunderstandings and disagreements would ultimately lead Spain towards the Civil War.

The *National Institute*, directed also from 1910 by Santiago Ramón y Cajal, incorporated several pre-existing laboratories, such as the Spanish Nobel laureate's Laboratory of Biological Research, the Museum of Natural Sciences, the Royal Botanical Garden and the Museum of Anthropology. But also new dependent institutes were created for what we have called elsewhere the 'first generation JAE' (González Redondo and Fernández Terán, 2004). They included: the Laboratory of Physical Research, for Blas Cabrera; the Mathematical Laboratory and Seminar, for Julio Rey Pastor; the Laboratory of Physiology, for Juan Negrín López; the Laboratory of Biological Chemistry, for Antonio Madinaveitia Tabuyo. Through these laboratories, during the second decade of the century, Spain's new system for higher education and research was able to collect certain scientists in its own centres. These were scientists who, for the first time in history, enjoyed a national fellowship programme through which they learned research skills with the most outstanding European figures in Germany, France, Switzerland, etc.

In 1923 a military coup led by Miguel Primo de Rivera developed into a dictatorship but, although there were some attempts to impose control on intellectual activities, in fact it was the same academics and scientists who were to continue directing teaching and research in the existing institutions. Moreover, in 1926 the Rockefeller Foundation (USA) decided to fund, in Madrid, what was meant to be one of the most advanced research laboratories in Europe, the Board's National Institute of Physics and Chemistry, which evolved from the Laboratory of Physical Research led by Blas Cabrera (Glick, 2005). And, in those same years, in order to balance the official support between the humanities and sciences, the dictatorship financed the acquisition of new buildings for the Centre for Historical Research.

Enrique Moles, with the collaboration of other chemists as Ángel del Campo and Obdulio Fernández, and the support of the dictatorship, organised the 9th Congress of the International Union of Pure and Applied Chemistry. It was the first Union Congress to be celebrated after World War I that would welcome German scientists once their country was admitted as a member of the Union in 1929.

Although Ramón y Cajal retired in 1922, the field of biomedical research continued to develop during the monarchy of Alfonso XIII, approaching European levels of activity. This was especially the case in the centre that would bear Cajal's name and be inaugurated in 1932, and through some of Cajal's most talented disciples such as Jorge F. Tello Muñoz, Fernando Lorente de No and Fernando de Castro. Although the Cajal Institute would also integrate the former Brain Physiology Laboratory, other scientists as Nicolás Achúcarro, Director of the Laboratory of Normal and Pathological Histology and his successor, Pío del Río Hortega, would also lead the process of convergence with Europe. In particular, Pío del Río Hortega would enter the National Institute of Oncology in 1928, becoming its director in 1932, and would at once rename it the National Cancer Institute (García Barreno, 2004).

On the other hand, and until the advent of the Republic, the intellectual elite barely became involved in putting an end to widespread illiteracy. In fact, it was the Provisional Government who had to undertake the construction of thousands of small schools in remote villages and who launched initiatives such as the Educational Missions. These brought culture to rural areas where theatre, cinema, art exhibitions and their ilk had never been experienced before (Otero Urtaza, 2006).

But the arrival of Fernando de los Ríos at the Ministry in December 1931 would mark a golden age for those intellectuals gathered around the centres of the Board. Thus, in addition to expanding the experience of Madrid's Experimental High School to Barcelona, Valencia and Seville, in August 1932, the Minister announced a completely new educational experience:

> We are going to create the aristocracy of the spirit ... The International University of Santander, nourished with Spanish and foreign teachers, with two student fellows selected and supported from each institution of higher education and each university, provided that, given the position that this new regime gives to culture, those two will be selected because of their competence and not because of their economic potential (*El Cantábrico*, 2 August 1932).

And indeed, Santander after the summer of 1933 would become a focus for science in Europe, directed by professors from the Board such as Ramón Menéndez Pidal and Blas Cabrera, and hosting such meetings as the Conference on Chemistry that would serve as a prologue to the 9th Congress of the International Union to be held in Madrid in 1934.

But the summit for the process of convergence with Europe was obtained through the National Institute of Physics and Chemistry, once the Republican government had assumed the compromises acquired with the Rockefeller Foundation during Primo de Rivera's dictatorship. Since the day of its official opening on 6 February 1932, several European scientists began to come to the Institute in order to undertake research under the direction of such leading Spanish professors of physics and chemistry as Enrique Moles, Julio Palacios, Miguel Catalán and Antonio Madinaveitia. All of these scientists had, by that time, achieved international status coordinated by Blas Cabrera. Since then, Spain would not just send scientists to learn abroad, but also welcome European and Spanish-American researchers in its own centres. So, at least in some fields, the situation felt at the end of the nineteenth century had almost been restored.

In fact, during the Republic the work of the 'first generation JAE' could finally correspond with the effort made in Spain through the Board, providing a research-based reform of Spanish higher education. Under their tutelage and guidance, and due to the retirement of several professors together with the creation of new chairs, from 1933 onwards, the 'second generation JAE' entered Spanish universities. This generation was the first to be properly trained in Spain that could match European standards. Amongst the members of this second generation we can find chemists such as Miguel A. Catalán, physicists like Arturo Duperier, mathematicians such as José Barinaga, and such geneticists as Antonio de Zulueta.

But, since its creation in 1907, the JAE had focused its attention on pure science, disregarding, for example, possible applications to industry and agriculture. Also the Board had not been able to cover some areas that were unavoidable for the Republic, such as the decentralisation of scientific research based exclusively in Madrid and the attraction of funding from private companies to centres that seemed to be only academic. Consequently, a new Republican institution was conceived to fill the gap: the National Foundation for Scientific Research and Trials of Reform (*Fundación Nacional para Investigaciones Científicas y Ensayos de Reformas* – FNICER). It was chaired by José Castillejo, who had to resign from the position he held at the Board.

But this new Spain, that had germinated from the ashes of the *1898 disaster*, did not satisfy everyone. For example, in 1940, Francoist Antonio de Gregorio Rocasolano summarised the process of convergence from a completely different perspective:

> The Free Institution for Teaching, who quietly, following its accustomed tactic, had become sole owner of the Board for Advanced Studies and Scientific Research, its laboratories and other official departments, freely controlled their destinies and, as originally had pretended, managed to locate only in Madrid, under its own auspices, the whole of the officially supported and financed research, hogging its more or less authentic achievements (Gregorio Rocasolano, 1940, p. 149).

Science and Culture during the Spanish Civil War

Since the Popular Front, the alliance of almost all republican and left-wing political parties, had won the elections in February 1936, everyone knew that Spain was being led to that 'tragic situation' that came into being through the military plot that caused the Civil War in July 1936. On one side were the Army and extreme right-wing groups of fascist falangists, traditionalists and so on, who soon found the support of conservative Spain. On the other, were moderate republicans who wanted democratic reforms together with those who longed for a socialist, communist or anarchist revolution. In short, Spain was facing a struggle between those who stuck to tradition and those who dreamed of change.

Indeed, if the Republic in peace manifested a definite willingness about supporting scientific activities and educational reforms, its commitment following the outbreak of the Civil War can be seen as surprising. This was made explicit from Valencia, where the Government led by socialist Francisco Largo Caballero had moved in November 1936 as Franco's troops were approaching Madrid. As early as January 1937, when communist Jesús Hernández Tomás, Minister of Public Instruction (several months later also Minister of Health), decided that: 'all scientific activities and works should continue or be resumed even with greater intensity than to the extent permitted under the current circumstances; giving preference, as it seems natural, to those works that may have a direct or indirect application to the needs of war' (*Proceedings* of the JAE, 13 January 1937, p. 8).

This support for scientific activities materialised especially in those centres run by the double Ministry in Valencia, Barcelona and even in besieged Madrid. Thus, in the capital, the synthesis of nicotinic acid to prevent niacin deficiency (a pandemic disease during the war) was achieved by the team led by Ángel del Campo in the National Institute of Food Hygiene [*Instituto Nacional de Higiene de la Alimentación*]. Also in Madrid, investigations continued in the National Institute of Physics and Chemistry directed by Enrique Moles, who also held the post of General Manager of Powders and Explosives for the Ministry of War in Barcelona. Even purely mathematical research, with no practical application and of no help to war efforts was encouraged and financed at the JAE's Mathematical Laboratory, even until the final stages of the conflict.

In fact, in Republican Spain during the war, for the first time in Spanish history, the working classes could come into contact with scientific knowledge previously reserved for teaching to the children of the bourgeoisie. On the one hand, the Militia of Culture [*Milicias de la Cultura*] would bring literacy and culture to workers fighting in the trenches. On the other, the Workers' High Schools [*Institutos para Obreros*], proletarian reincarnations of the once elitist Students Residence and the Experimental High School of the JAE would provide intensive courses to those teenagers who were supposed to be prepared for ruling the country once the war had finished.

But if the Republic made a big effort in preserving scientific activities during the war, a significant group of intellectuals corresponded through successive calls to the international community in the form of manifestos published in the Spanish press and sent to European institutions (González Redondo, 2013). They were the members of the so called House of Culture [*Casa de la Cultura*] organised in Valencia with relevant professors of Madrid's University and members from the Board's centres, including: Pedro Carrasco Garrorena, Arturo Duperier, Antonio Madinaveitia, Manuel Márquez, Enrique Moles, José Puche, Gonzalo Rodríguez Lafora, José M. Sacristán, Jorge F. Tello and Antonio de Zulueta.

But not all intellectuals remained loyal to the legitimate rulers, like the so-called 'Third Spain', the group of professors who did not compromise with any of the two sides in conflict (González Redondo and Fernández Terán, 2010). In fact, on 2 December 1937, the Minister denounced them as: 'a group of academics, showing a clear lack of solidarity with the Spanish people, who fight defending national freedom, who have openly failed to fulfil their most basic duties, ignoring the call of the government' (GR, 4 December 1937, pp. 904–5). Thus, the Republic was forced to expel relevant figures from their professorial chairs who had chosen to continue their careers and research, since the beginning of the Civil War, in countries like France, United Kingdom, USA, Argentina and so on. Among them we find the names of Blas Cabrera, Américo Castro, Hugo Obermaier, José Ortega y Gasset, Luis Recasens, Claudio Sánchez Albornoz, Xavier Zubiri and Luis de Zulueta.

In contrast to the spirit of such punitive order, several scientists enjoyed permission from the Republican government to move abroad during the war and continue their research in peace. Examples include physiologist Severo Ochoa as early as September 1936, histologist Pío del Río Hortega in January 1937, neurologist Gonzalo Rodríguez Lafora in March 1938 and physicist Arturo Duperier in April 1938.

Purge and Exile in Spanish Science

As the Civil War approached an end, Francoist Spain intensified its reorganisation and so also the purge, not only of all scientists committed to the Republic, but also of those who had remained neutral. Thus two Orders signed by the Minister of National Education on 4 February 1939 expelled a significant number of professors from universities because 'it was open and notorious' their 'disaffection to the new regime introduced in Spain, not only for their behaviour in areas that have suffered or are still suffering the Marxist domination, but also for their persistent anti-national and anti-Spanish policies in the times preceding the Glorious National Movement' (BOE, 7 February 1939, p. 932).

Among the scientists who were sanctioned we find Republican ministers such as José Giral, Fernando de los Ríos and Juan Negrín, also personalities who held positions in the Republican administration including José Castillejo, Manuel Márquez, Honorato de Castro and Cándido Bolívar. Additionally, amongst those sanctioned were scientists committed to the Republic at war, such as Enrique Moles, Pedro Carrasco and Antonio Madinaveitia, and also early disaffected scientists such as Blas Cabrera, Teófilo Hernando and Gustavo Pittaluga.

But in order to clarify its purpose, the second Order noted: 'The evidence of the pernicious behaviour for the nation of those professors just mentioned makes all those procedural safeguards, which would otherwise constitute the fundamental condition for any prosecution, totally useless' (BOE, 17 February 1939, p. 932). As such, Francoist Spain thwarted any legal provision, in particular the provision signed by Franco on 11 November 1936 at the beginning of the war, which meant the starting point of the purge process introduced as described above. Indeed, the new Regime was determined 'to remove from its roots through this process those false doctrines that, together with its apostles, have been the main factors for the tragic situation our Nation has been led to' (BOE, 11 November 1936, p. 153). And, if necessary, they would not even obey their own stipulations.

But the process had just started. A new law signed on 10 February 1939 and published in the *Official Gazette* established the 'Rules for the purge of all civil servants':

> It is the wish of the Government to carry out this purge with maximum swiftness and within flexible rules that will allow the fast return to their positions of those civil servants which deserve it for their background and behaviour and, at the same time, to impose appropriate sanctions, according to the cases, to those who having failed to perform their duties, contributed to the subversion and gave unforgivable assistance to those who by violence and regardless of any legal provision, seized the leading post of public Administration (BOE, 14 February 1939, p. 856).

It was intended that those people committed to the new Regime would be reincorporated as soon as possible, so they could serve as witnesses in subsequent prosecutions of those suffering the purge. But everyone got involved, some as prosecutors, others as defendants. No one was safe, no one could remain neutral. Besides a personal affidavit, they were all forced to make successive statements incriminating colleagues about whom there was uncertainty, suspicion or where there was simply a desire to show that the Regime could enforce any of the accusations made possible by the law.

As might be expected, many professors that remained loyal to the legitimate Republican government, and who certainly had shared 'dissolving ideologies and institutions, in open opposition to the national wit and tradition' (BOE, 11 November 1936, p. 153) decided to go into exile whenever the circumstances allowed them to do so. The case of the Physics section at Madrid's Faculty of Science is highly significant, where only Julio Palacios retained his chair while all his colleagues had been forced to leave Spain or, like Miguel A. Catalán, were suffering what has been characterised as an internal exile in their own country.

On the other hand, the defeat of Republican Spain would encourage the return of right-wing mathematicians from Argentina, such as Esteban Terradas, who had left revolutionary Barcelona in early 1937, or Julio Rey Pastor, who had lost his chair at Madrid's University in the 1930s and was already living in Buenos Aires before the war. In contrast to what happened to many physicists and chemists, most mathematicians retained their chairs under the new Spanish Regime, although the Director of the JAE's Mathematical Laboratory, José Barinaga, also suffered his particular internal exile until 1945. Nevertheless, before his final return to Spain, Rey Pastor welcomed and helped several exiled mathematicians in Argentina, such as Manuel Balanzat, Miguel A. Santaló, Ernesto Corminas, Pedro Pí Calleja and, as late as 1945, Francisco Vera.

In any case, the most common first stop for exiled Spanish scientists was France, both individually and as a collective. There, and as early as February 1939, the still Prime Minister Juan Negrín used part of the funds deposited in foreign banks to create the Service for the Evacuation of Spanish Refugees [*Servicio de Evacuación de Refugiados Españoles* – SERE]. But the disputes that had conditioned the entirety of Republican politics during the war (especially in the Socialist party, but also between republicans, communists, anarchists and so on) continued after their defeat. Thus, in late July, the former socialist Minister Indalecio Prieto, on behalf of the Permanent Delegation of the Republican Parliament in exile, created the Board of Aid to the Spanish Republicans [*Junta de Auxilio a los Republicanos Españoles* – JARE]. Initially both institutions established their headquarters in Paris, as did the first organisation of exiled professors in December 1939, the Union of Spanish University Professors Abroad [*Unión de Profesores Universitarios Españoles en el Extranjero* – UPUEE], chaired by Gustavo Pittaluga.

From a personal perspective, I have already mentioned Pío del Río Hortega, who had been proposed for the Nobel Prize for Physiology in 1929 and 1934, and who continued his work in Paris before moving to the United Kingdom and finally to Argentina, where he died in 1945 (López Piñero, 1990). Duperier also moved from Paris to Manchester and then to London, where he became a leading figure in cosmic ray research and could also have been proposed for the Nobel Prize for Physics in the 1950s if he had renounced his Spanish nationality (González de Posada and Brú Villaseca, 1996). Physicists such as Manuel Martínez Risco or chemists like Enrique Moles entered the French National Centre for Scientific Research [*Centre National de la Recherche Scientifique*] and could resume their studies as *Maîtres de Recherches* in Paris.

But the greatest aid came from Mexico, where Spanish refugees had been hosted since 1937. At the end of the war, Mexican President Lázaro Cárdenas commissioned Salvador Zubirán and Ambassador Narciso Bassols to meet representatives of SERE, JARE and UPUEE in order to 'select the best among republican refugees concentrated in French camps' so that they should constitute 'the best immigration for Mexico' (Piña Soria, 1939, p. 12). These refugees would have to reconcile their activities according to the needs of Mexican economic and social development. To achieve this goal, Cárdenas ordered 'a careful selection ignoring any political or social affiliation from the immigrants' (Piña Soria, 1939, p. 13).

The selection was made in France by the House of Spain [*Casa de España*] initially through the SERE, and organised in Mexico by the SERE's Technical Committee to Aid Spanish Refugees, from where cultural and educational institutions including the Luis Vives Institute, the Cervantes Patronage, Séneca Publishing Company, and others were newly created in order to integrate the refugees into the life of the country. But, by July 1939, the SERE had exhausted its resources and would be replaced by the JARE, from where new institutions as the Spanish-Mexican Academy and the Spanish-Mexican Institute Pedro Ruiz de Alarcón were also created.

The House of Spain, renamed College of Mexico [*Colegio de México*] in October 1940, based its decisions and selections on the network of contacts established during the decades prior to the Civil War. These came primarily from the JAE, which in 1910 had also been assigned the management of foreign academic exchanges and had been in intimate contact, since 1926, with the Spanish-Mexican Institute for University Exchanges (Dosil Mancilla, 2010). Again, the group of exiled intellectuals gathered in Mexico (and many other American countries) shared a scientific background, that of the JAE, and a common *spirit*, that of the old Free Institution for Teaching. This *spirit* would permeate all those newly created educational centres just mentioned and several others established later on by the College of Mexico.

But the initiative that best illustrates the aspiration of the exiled scientists to build a common identity, and that would characterise them as a group before both the host countries and the international community, was the journal *Ciencia* [Science] published in Mexico between 1940 and 1975. This served as a platform for the scientific work of those Spanish scientists exiled in over twenty different countries. In short, the wide gap between the *two Spains* would continue for several decades.

Final Remarks: Towards Reconciliation

It is true that several scientists were allowed to return to Spain after the end of the Civil War, but they were more the exception than the rule. Nevertheless, by 1970 things were changing as the Regime was approaching its end. Young Spanish researchers once again enjoyed scholarships in Europe and the USA. Even exiled scientists such as Nicolás Cabrera (Blas Cabrera's younger son) accepted the invitation of Franco's Minister of Education in 1970 to establish the Department of Physics of the newly created Autonomous University of Madrid [*Universidad Autónoma de Madrid*]. And an analogous proposal was accepted by Severo Ochoa (Nobel Prize for Physiology in 1959, but an American citizen) who also established the Institute of Molecular Biology in Madrid in the early 1970s.

Indeed, as soon as the Dictator died in November 1975, the Spanish people headed towards the restoration of democracy. The Amnesty Law, promulgated on 15 October 1977, in its first article amnestied all acts with political intentions, whatever the outcome, that had been defined as offences and carried out before 15 December 1976. In its seventh article it also provided that all sanctioned civil servants should be reintegrated in the fullness of their active and passive rights and could return to the same positions where they had lost them. It also eliminated all possible criminal records and unfavourable notes in their personal files, even if the offender had died.

Finally, the promulgation of the new Constitution in late 1978 would put an end to that Spain of the 'national wit and tradition'. After recouping their chairs in 1977, Francisco Giral and Augusto Pérez-Vitoria witnessed that Spain was prepared for definite reconciliation.

It is obvious that Franco had won successive battles between 1936 and 1939, but I firmly believe that Francoism had begun to lose the war from 1939, precisely because of his decision to deter any project of national reconciliation. In other words, because of his determination to condemn the Spanish to the division between winners and losers, between the guardians of 'tradition' in the homeland and those who had dreamed to modernise and Europeanise Spain, the losers, exiled abroad or confined in their own country. Reconciliation was possible. Franco had been finally defeated.

List of References

Berrojo, R., 1980. *Enrique Moles y su obra*. 3 Vols. Barcelona: Universidad de Barcelona

Costa, J., 1906. 'Preface'. In R. Sánchez Díaz, *Juan Corazón*, pp. vii–xxi. Madrid: Librería de Fernando Fe.

Dosil Mancilla, F.J., 2010. 'La dinámica de las redes del exilio científico en México'. In: J.L. Barona (ed.), *El exilio científico republicano*. Valencia: PUV, pp. 249–62.

García Barreno, P., 2004. 'Panorama de las Ciencias de la Vida en España en la época de Cajal'. In: F. González de Posada et al. (eds). *Actas del II Simposio Ciencia y Técnica en España de 1898 a 1945: Cabrera, Cajal, Torres Quevedo*. Madrid: Amigos de la Cultura Científica, pp. 143–96.

Giral González, F., 1994. *Ciencia española en el exilio, 1939–1989. El exilio de la ciencia española*. Barcelona: Anthropos.

Glick, T.F. et al. (eds), 2001. *The Reception of Darwinism in the Iberian World*. Dordrecht: Kluwer.

Glick, T.F., 2005. 'Dictating the Dictator: Augustus Trowbridge, the Rockefeller Foundation and Physics in Spain, 1923–1927'. *Minerva*, 43, pp. 121–45.

González de Posada, F., 2009. *José Celestino Mutis: otra perspectiva científica con el trasfondo de Jorge Juan*. Madrid: Fundación Jorge Juan.

González de Posada, F. and Brú Villaseca, L., 1996. *Arturo Duperier: mártir y mito de la ciencia española*. Ávila: Institución Gran Duque de Alba.

González Redondo, F.A., 2004. 'El panorama de la ciencia española entre 1898 y 1945'. In: F. González de Posada et al. (eds). *Actas del III Simposio Ciencia y Técnica en España de 1898 a 1945: Cabrera, Cajal, Torres Quevedo*. Madrid: Amigos de la Cultura Científica, pp. 11–34.

González Redondo, F.A., 2013. 'Los científicos españoles entre la República *en paz* y la República *en guerra*: compromisos y desafecciones'. In: A. Gómez Rodríguez and A.F. Canales Serrano (eds). *Estudios políticos de la ciencia. Políticas y desarrollo científico en el siglo XX*. Madrid: Plaza y Valdés, pp. 53–80.

González Redondo, F.A. and Fernández Terán, R.E., 2002. 'Nuevas perspectivas en torno a la política de pensiones de la Junta para Ampliación de Estudios: modelos de encuentro con Europa de la Universidad española'. *Revista Complutense de Educación*, 13(2), pp. 563–93.

González Redondo, F.A. and Fernández Terán, R.E., 2004. 'El criterio de relevancia científica y la organización histórica por generaciones de la ciencia española'. *Revista Complutense de Educación*, 15(2), pp. 687–700.

González Redondo, F.A. and Fernández Terán, R.E., 2010. 'La tragedia de la Tercera España: el exilio de Blas Cabrera'. In: J.L. Barona (ed.). *El exilio científico republicano*. Valencia: PUV, pp. 89–109.

Gregorio Rocasolano, A., 1940. 'La investigación científica, acaparada y estropeada'. In: F. Martín-Sánchez et al. *Una poderosa fuerza secreta. La Institución Libre de Enseñanza*. San Sebastián: Editorial Española, pp. 149–60.

Jiménez-Landi, A., 1996. *La Institución Libre de Enseñanza y su ambiente*. 4 Vols. Madrid: Editorial Complutense.

Juan, J., 1774. *Historia de la Astronomía en Europa*. Madrid: Imprenta de la Gazeta.

López Piñero, J.M., 1990. *Pío del Río Hortega*. Madrid: Fundación Banco Exterior.

López Sánchez, J.M., 2006. *Heterodoxos españoles. El Centro de Estudios Históricos, 1910–1936*. Madrid: Marcial Pons.

Martín-Sánchez, F. et al., 1940. *Una poderosa fuerza secreta. La Institución Libre de Enseñanza*. San Sebastián: Editorial Española.

Morodo, R., 1985. *Los orígenes ideológicos del franquismo: Acción Española*. Madrid: Alianza Editorial.

d'Ocagne, M., 1930. 'Machines a Calculer'. *Figaro*, 25 May 1930, p. 5.

Otero Urtaza, E. (ed.), 2006. *Las Misiones Pedagógicas, 1931–1936*. Madrid: Publicaciones de la Residencia de Estudiantes.

Piña Soria, A., 1939. *El Presidente Cárdenas y la inmigración de españoles republicanos*. Mexico: Multígrafos SCOP.

Pratt, D.J., 2001. *Signs of science, literature, science and Spanish Modernity since 1868*. West Lafayette, IN: Purdue University Press.

Silió, C., 1933. *En torno a una revolución*. Madrid: Espasa-Calpe.

Unamuno, M., 1895. 'En torno al casticismo. Sobre el marasmo actual de España'. *La España Moderna*, 78, pp. 26–45.

Serial References:

BOE – *Boletín Oficial del Estado*
GM – *Gaceta de Madrid*
GR – *Gaceta de la República: Diario Oficial*

Chapter 4

The Reactionary Utopia: the CSIC and Spanish Imperial Science[1]

Antonio Fco. Canales

The CSIC from a Historiographical Perspective

The Higher Council for Scientific Research (*Consejo Superior de Investigaciones Científicas* – CSIC) is a state institution that was set up immediately after the end of the Spanish Civil War by Franco's government. The CSIC was established to replace the Board for Advanced Studies and Scientific Research (*Junta para la Ampliación de Estudios e Investigaciones Científicas* – JAE), the institution that had been promoting scientific research in Spain since 1907, and whose staff and facilities were now taken over by the newly-created Higher Council. Despite the key role played by the CSIC in the scientific work carried out during the early years of Franco's Regime, the historiography on Spanish science offers no global interpretation of this new institution. In the absence of such an interpretation, the studies that do analyse the institution oscillate between two implicit interpretative approaches. On the one hand, some studies tend to consider the CSIC as a mere continuation of the JAE, the institution it was established to replace; and, on the other, it is seen as a manifestation of the Regime's autarkic policies.

Those advocating the first approach tend to believe that, despite its new management, the aims of the CSIC were similar to those of the JAE, namely to ensure the upkeep of research centres, establish research priorities and award foreign study grants. This interpretation supports those views of the CSIC that, for one reason or another, prefer to gloss over or underplay the importance of the rupture occasioned by the Civil War. Examples include the institutional

---

[1]   This article was made possible thanks to the Spanish National Research Projects FFI2009–09483 (Ministry of Science and Innovation) and FFI2012–33998 (Ministry of Economy and Competitiveness) and the Research Projects of the Universidad de La Laguna, 2012 and 2013.

publications of the CSIC itself (Puig-Samper, 2007) or the work of the hagiographers of the first leader of the new institution (Gutiérrez, 1990).

This interpretation, which underscores the continuity of the institutions before and after the Civil War, is problematic for a number of different reasons. Firstly, the basic premise of the CSIC was precisely the radical rejection of the JAE and all that it stood for. The CSIC stemmed from an explicit desire to completely break away from Spain's scientific past and start afresh. The scientific communities of the pre-war period were dismantled by means of a harsh process of purge and exile, the JAE was dissolved and an attempt was even made to physically delete all traces of its existence through the building of a new campus over its old facilities. Few events are more symbolic of this process of eradication than the conversion of the Auditorium of the Students' Residence into a church in order to expiate the sins committed by the Spanish scientific community. Secondly, the aims of the JAE and the CSIC were diametrically opposed. Despite its multipurpose nature, the JAE was based on one very clear objective: to bring Spain into Europe as a means of redressing the country's scientific and cultural backwardness. As it will be shown below, the viewpoint espoused by the CSIC was completely the opposite: it questioned Western science and its concept of *scientific backwardness* and aspired to return to traditional Spanish science, the science of the Empire prior to the Western scientific revolution. Finally, the CSIC dissolved the social contract for science embodied by the JAE and substituted it with the categorisation of scientists under the management and control of the New State (Gómez and Balmer, 2013 and Gómez, 2014). Although it is true that from the 1950s onwards the similarities between the CSIC and the JAE increased, the time dimension here is important to a correct reading of the nature and aspirations of the new Higher Council for Scientific Research. A failure to take this time dimension into account glosses over the initial objectives of the CSIC, reinterpreting them in accordance with what the institution became later on in its development. From a political perspective, underplaying the radical rupture between the two institutions is also a clear attempt to appropriate the well-regarded and highly valued legacy of the JAE by those who were responsible for its dissolution.

The second interpretation of the CSIC, which underpins studies of Spanish science during this period, views the new institution as an instrument for placing scientific research at the service of the autarky, the nationalist and interventionist economic policy which reigned supreme during the first two decades of Franco's rule. This view holds that the CSIC was established in response to the need to produce raw material and energy sources, as well as technology, in order to reduce Spain's dependence on the outside world as much as possible – a priority objective of the Regime's autarkic policy. This interpretation is supported by studies on the applied research carried out at the CSIC (López, 1997, 1998 and 2008). As will be shown, it does not seem possible to deny this autarkic dimension of the CSIC, but this dimension alone is not sufficient to characterise the institution for (at least) two reasons. Firstly, the section responsible for applied research, the Juan de la Cierva Centre, was only one part of the CSIC, albeit an important and well-funded one. We should not lose sight of the fact that the structure of the CSIC was a globalising one, with six centres and over forty research institutes, the majority of which could do very little to satisfy the material needs of autarky. Furthermore, the time dimension is important here once again. The prioritisation of production for autarky only became important from 1944 onwards, when it became clear that the Allied forces were going to win the Second World War. Nevertheless, the CSIC was set up at the end of 1939. During this space of time, research at the service of autarky was not deemed a priority, since the Regime considered itself to be an ally of the most technologically-powerful nation in the world: Nazi Germany (López, 2008, p. 97). It was only when it became clear that the Fascists were going to lose the war and Franco's Regime began preparing to isolate itself from the new international order of the post-war period that an urgent need was felt to encourage the research required to make this step possible. In the words of Santiago López García (1998, p. 11), 'for the first time in Franco's rule, research became a necessity rather than an excuse for spreading state propaganda'. Even so, the autarkic aim became considerably more relaxed from 1953 onwards, when Franco finally agreed to put an end to international isolation and access American patents.

Neither of the two interpretations outlined above seem to pay any serious attention to the reasons publicly espoused by Spain's political and scientific authorities for the creation of the CSIC. The members of Franco's Regime who were responsible for science ceaselessly alluded to aims such as a return to imperial science or the overcoming of the dichotomy between faith and reason that arose during the seventeenth century as a result of the scientific revolution. Historiography seems to underplay these aspirations, reducing them to a single, unimportant example of the mystical and imperial rhetoric of the era, a kind of unrealisable dream fed by the euphoria of victory that was simply just another anecdotal part of life during that particular period. But these aims are anything but unimportant. On the contrary, they should be taken very seriously, since they resulted in science being included in the coherent, extremely ambitious programme that strove to redefine all aspects of Spanish society: the National-Catholic project. In the same sense in which Michael Mann advocates that fascist ideology should be taken seriously (Mann, 2004, p. 3), in relation to Spain it is vital to take that stated by the National-Catholics very seriously indeed, despite the fact that their approaches may seem extemporaneous and unfeasible from a modern-day perspective.

The aim of this chapter is to analyse the approaches adopted by the men who designed the CSIC in order to offer a global interpretation of the new institution as a product of the National-Catholic project. First of all, two introductory sections will outline the basic traits of the National-Catholic project and the policy initially adopted by Franco's Regime in relation to science. Subsequently, the central section of the chapter will focus on the CSIC as a product of the reactionary utopian dream of returning to an imperial science subject to religion. To this end, the section will describe the *new Spanish science* to which the Regime aspired, based on an analysis of the speeches given by the Minister of Education, and will show the extent to which the structure of the CSIC depended on this view of science. It will also chart the underlying desire to control both knowledge and academic and scientific communities. Finally, the chapter will describe the failure of the project and will chart how its advocates sought to establish a pragmatic authoritarian compromise with the country's scientific communities.

National-Catholicism

National-Catholicism was the ideological framework which dominated Franco's dictatorial regime. It was an ideological merging of reactionary fundamentalist thought and modern fascist formulas of political and social dominance. Its basic premise was the consubstantiality of Spain and Catholicism, which turned Catholicism into the core of the Spanish nation, the backbone of national history and, in short, the very essence of Spain.

National-Catholicism was based on an account of history that underscored the identification of national unity with religious unity. Spain had been forged during the Middle Ages in a centuries-long national and religious struggle to expel the Muslims from the Iberian Peninsula. The political unification of the country under the Catholic Monarchs at the end of the fifteenth century coincided with its religious unification following the defeat of the last Muslim kingdom and the expulsion of the Jews. Furthermore, during the sixteenth century, the Empire had conferred an international providential mission on this synthesis of nation and religion. For the National-Catholics, the empire had not been a territorial or political undertaking, but rather a religious enterprise. Imperial Spain had fulfilled its providential mission to evangelise America and to take up arms to defend Catholic dogma in Europe against the Protestant threat.

The Empire, the period during which Spain had imposed on the world the fusion of Sword and Cross, represented for National-Catholicism the fullest, most genuine expression of the Spanish nation. This glorious period had been followed, from the sixteen-hundreds onwards, by centuries of decline. Significantly, this decline coincided with Europe's advance towards modernity through scientific and technical development, although mainly through an intellectual revolution based on rationalism, empiricism and materialism, all principles to which National-Catholicism was virulently opposed. Spain thus became the guardian of the true universal values, which were erected in opposition to the erroneous path of intellect which prevailed in the Western world. According to this view, the basic error committed by drifting European thought, which was dragging Western civilisation inexorably towards a major crisis (as proven indeed by the Second World War), was its questioning of the supremacy of medieval Catholic dogma. From a National-Catholic perspective, Protestantism led to communism, passing first through the Enlightenment, liberalism and socialism. It was no coincidence that José Pemartín, one of the leading National-Catholic theorists and the head of secondary and higher education under Franco's Regime, had no qualms in stating that 'communism was born in Eisleben with Luther' (Pemartín, 1937, p. 46). Standing firm against Europe's headlong dive into suicidal chaos, Spain saw itself as the 'spiritual redoubt of the West', a slogan that was fiercely exploited by Franco's Regime right up until the very end.

From this perspective, Spain's decline was viewed not only as a loss of political and military clout, but also, and most importantly, as a process of internal denaturalisation. The decline was first and foremost the result of Spain's abandonment of the values that constituted its very essence; in other words, it was caused by the country's denial of itself and acceptance of foreign, anti-Spanish ideas. In contrast to this process of national degeneration, the Civil War was seen as a heroic exploit seeking to enable the rebirth of the true Spain. The basic premise of this regeneration was, naturally, the elimination of all those agents which had introduced wrongful, pernicious foreign ideas into the country; in short, it was the elimination of the anti-Spain. Repression took diverse forms and while it was mainly directed against communists, anarchists, socialists and democrats, it was also wielded against liberals, who were seen as having sown the seeds of the dissipative ideologies that gave rise to all the rest. Hence the emergence of the idea that intellectuals had a special responsibility;

an idea which made up the backbone of Franco's policies in relation to culture and science.

The First Measures Taken in Relation to Science

The National-Catholic convictions of the new regime were bound to have an effect on science, since Spanish scientific development was (necessarily) based precisely on those insidious foreign ideas that had denaturalised the true Spain. A correct assessment of the effects of the National-Catholic victory over Spanish science first requires a clear distinction to be made between two levels or spheres of action. The first affected the specific process of Spanish scientific development, that is, the scientific communities and their institutions from before the war. The second, which was much more ambitious, diffuse and complex, was linked to the desire to establish a new kind of Spanish science in keeping with the National-Catholic principles of a return to the Empire.

The Settling of Old Scores with Pre-War Scientific Development

In relation to the first level, as has been shown in previous chapters, liberal, democratic Spain's main institution for the development of science had been the Board for Advanced Studies and Scientific Research (JAE). The JAE was a State institution attached to the Ministry of Education, which since 1907 had awarded grants for studying abroad and had set up many scientific research institutes. The JAE was strongly influenced by the Free Institution for Teaching (*Institución Libre de Enseñanza* – ILE), an association of Krausist liberals who aspired to regenerating Spanish society through education and science. Right from its foundation, which was a direct reaction to the expulsion of a group of university professors for defending academic freedom against enforced submission to Catholic dogma in 1876, the ILE had been the target of hostility from Catholics, traditionalists and, to a large extent, conservatives in general. The Spanish Civil War provided an opportunity for settling old scores with this influential intellectual movement and for putting an end to a long ideological battle that had been raging since the middle of the nineteenth century.

The ILE therefore became an obsessive reference in the discourse used by the rebels in relation to science. The association was presented as a secret sect that operated behind the scenes, pulling invisible strings of power with the sole aim of corrupting the academic and intellectual elite in order to destroy Spain. The conclusion drawn by advocates of this view was, of course, that Spain's entire academic and cultural sphere (that is, the people responsible for the Spanish revolution) was in urgent need of a thorough cleansing. In the words of the paediatrics professor Enrique Suñer Ordóñez, 'we need to pull our enemies out by their roots, removing first these intellectuals who brought about the catastrophe. Since they are the most intelligent and the most cultured, their responsibility is that much greater' (Suñer, 1937, p. 201).

The purging of the world of thinking, culture and science became one of the new regime's top priorities. The aim was to eliminate all those who, over the years, had been propagating foreign, anti-Catholic ideas from their seats at the universities and research centres. As Minister José Ibáñez Martín stated, 'it was therefore vital for our culture to amputate the corrupt limbs with force and energy, to cut away the rot with sure, implacable strokes of the scythe and to cleanse and purify all harmful elements'. This inquisitorial zeal should also remain inflexible and resolute in the face of 'falsely humane considerations' (Ibáñez, 1940b, p. 10).

The antagonism of the victors towards the ILE spilt over to the JAE also. In the eyes of the new regime, the links which existed between the JAE and the men of the ILE rendered the whole policy of cultural and scientific renovation and modernisation which had been developed throughout the first third of the century by this public institution a mere sectarian strategy designed to beguile and contaminate Spain's cultural and scientific elite and younger generations. From this standpoint, there was no option but to wipe the slate clean and eliminate all traces of the science policy of the pre-war period by dismantling its principal institution. In the words of the Minister: 'We must dismantle the whole set up of a false high culture that distorted the national spirit with division and discord, and eradicate it from the spiritual life of the country, cutting off its tentacles and closing off all avenues of return' (Ibáñez, 1940b, p. 9). Consequently, the JAE was abolished and its scientists severely purged (López, 1997, p. 227). Scientists were also purged in the parallel process of purge of the universities (Otero Carvajal, 2006; Claret, 2006; Giral, 1994; Morente, 2005 and 2015). In general, the victory of Franco's forces resulted in the dismantling of the pre-war scientific communities through purging and exile.

The First Science Policy

The inquisitorial desire of the new regime to hold scientists accountable for past actions pushed the definition of a new science policy somewhat into the background. Nevertheless, sooner or later the victors had no choice but to tackle this rather knotty issue. Purging in accordance with the principles of National-

Catholicism posed little or no problem at a theoretical level; it was enough to apply these principles in a negative sense in order to remove from their post anyone who could not produce evidence of religious conviction and anti-liberal affiliation. However, it was a lot less simple to apply these same principles in a positive sense to science policy. The question was, what precisely would constitute a truly Catholic and Spanish science policy? Three responses to this question were proposed within the Regime throughout the course of 1939.

At the end of 1938, in an address delivered at the Athenaeum of Seville (which was then followed up by a series of newspaper articles and a pamphlet), the chemistry professor Manuel Lora Tamayo (1939) proposed a project to reorganise science following the war. Lora's project was based on the assumption of autarkic scientific nationalism, which had characterised Italian science during the interwar period (Maiocchi, 1993), and defended the need to direct scientific research towards the exploitation of national resources in order to reduce Spain's dependency on foreign imports. The aim was, in short, to prioritise applied research at the service of the autarky. To direct this broad-ranging project of autarkic technological development, Lora proposed the setting up of a National Research Council that was almost a direct copy of its Italian counterpart, and not only in name. Like the Italian *Consiglio Nazionale delle Ricerche*, the proposed council would not have its own facilities, but would rather coordinate and fund the applied research conducted in universities (Gómez and Canales, 2009).

Lora's proposal was not taken very seriously by Franco's first Minister of Education, Pedro Sáinz Rodríguez, who opted instead to include the management of science in the remit of the Institute of Spain (*Instituto de España*), a newly created body which encompassed the various Royal Academies and aspired to establishing and directing the new Spanish culture. The JAE and the National Foundation for Scientific Research and Trials of Reform (*Fundación Nacional para Investigaciones Científicas y Ensayos de Reformas* – FNICER) were dissolved and all their facilities and staff were transferred to the Institute of Spain. To fill the post of vice-president of the Institute (the person responsible for managing all these resources), the Regime appointed the only prestigious scientist with proven pro-Franco sympathies, Julio Palacios Martínez. Following the victory of Franco's troops in April 1939, this physics professor, who had himself trained at the JAE, was tasked with the job of reorganising the institutes and centres of the dissolved institutions and purging their remaining staff of any remaining undesirables (Malet 2009, pp. 317–18). In the opinion of A. Malet (2008), this first science policy was basically a policy of continuity, since it seemed to rely on the continuity of a thoroughly purged and totally reoriented JAE, under the management of scientists sympathetic to the new regime.

However, Julio Palacios hardly had time to start reorganising the laboratories and research centres before this first policy of continuation was radically cut short by the new Minister of Education, José Ibáñez Martín, who took possession in September 1939, just six months after Franco's troops had entered Madrid. Ibáñez proposed a radical alternative to both the institutional organisation and the directors of Spanish science. Julio Palacios was soon replaced as the person responsible for reorganising Spanish science by José María Albareda Herrera, a high school teacher with firm religious convictions and close ties to the incipient *Opus Dei*.[2] Despite the bonds of friendship which existed between Lora Tamayo and Albareda, the new team also failed to take the Seville professor's proposal seriously. Albareda and Ibáñez instead opted for a project that was more in keeping with National-Catholic proposals, and was much more radical in relation to science itself: the restoration of Spanish imperial science. The instrument established to turn this reactionary utopia into a reality was the CSIC, which was founded in November 1939.

The Reactionary Utopia: The CSIC and the Hierarchisation of Knowledge

The CSIC was much more than a body for reorganising scientific research following the purge and exile process; for its advocates, it represented the embryo of new Spanish science. This new science would abandon traditional scientific values and be ruled by a new aim, that of overcoming the clash between faith and reason that had existed since the scientific revolution. The objective was to undo the damage wrought by Western thinking over (at least) the last three centuries. The adjective *scientific* that was included in the title in relation to the term *research* should not be misinterpreted, since it did not refer to that which was traditionally understood by the word *science*, but rather to a new form of knowledge defined by its submission to God.

---

2   The *Opus Dei* is an organisation within the Catholic Church founded in Spain in 1928 by the priest José María Escriva de Balaguer, to whom Albareda was closely linked even to the point of escaping together from the Republican side during the Civil War through the Pyrenees. For the connections of Albareda with the early *Opus Dei*, see Camprubí 2014, pp. 44–7. The *Opus Dei* became an extremely powerful and influential group during the second part of the Franco Regime so that the internal political dynamics of the regime during its last two decades is usually explained in terms of competition between members of the *Opus Dei* and the falangists of the single party.

Spanish Science

Minister Ibáñez Martín left no room for doubt as to what the aim of the CSIC was to be, when he called, during the opening ceremony of the new institution, for a return to the Spanish science that was carried out during the imperial age (1940, p. 18), a concept that at other times he had referred to as the 'immortal science' or the 'true and eternal science' that Spain had taken to Europe and America (1940a, pp. 8 and 10). With this call for imperial science, the Minister explicitly aligned himself with the traditionalist sectors in the seemingly interminable controversy over Spanish science (García, 1970) against those scientists who had defended the need to bring Spain into the mainstream of Western science. Indeed, the Minister stated fairly bluntly that the creation of the CSIC put an end to the whole issue: 'that controversy ends today' (Ibáñez, 1940, p. 15). The CSIC, therefore, was based on an approach that was diametrically opposed to that of the JAE and the FNICER. It was not its aim to help bring Spanish research up to international scientific standards, but rather, on the contrary, to consolidate an alternative version of these standards through the restoration of another type of science – the Spanish imperial science.

This *science* implied concepts that were notably different from what was commonly understood by the term. To start with, Spanish imperial science wanted nothing to do with the rationalist, materialist thinking that had given rise to 'false European scientism' (Ibáñez, 1940a, p. 7). On the contrary, it repudiated the 'deification of reason' and was staunchly opposed to the 'foolish intellectual arrogance' of agnosticism (Ibáñez, 1940, p. 17). Hence, as the Minister stated in the opening ceremony of the first post-war academic year at the University of Madrid, Encyclopaedism was not considered a stimulus for Spanish science, but rather as the main cause of the crisis of 'our science', in addition to the full-on attack against the 'unshakable principles of our Homeland' (Ibáñez, 1939, p. 2). Spanish science was presented, then, as an alternative to European science and its main characteristic, in the pure National-Catholic logic, was its submission to God and acceptance of Catholic dogma. 'We thus conceive Spanish science as an effort by intelligence to comprehend the truth, as an aspiration towards Gods, as a philosophical unit, as the attainment of progress' (Ibáñez, 1940, p. 17).

From this perspective, the tension between faith and reason that had characterised the development of modern science was declared officially over: 'it is the science that will never be in conflict with faith' (Ibáñez, 1940, p. 17). For the Minister, this submission to religion called neither reason nor science into question; quite the opposite in fact, it merely brought Spanish science closer to *true* reason and *true* science. Here, Ibáñez established a distinction between the *true* principles and scientific values embodied by Spanish imperial science and the corrupted version that had won out in the West. Thus, Spanish imperial science involved a 'process of rational knowledge' and an understanding of the 'nature of scientific laws', as opposed to the 'scientific heresies' which dominated European science. In short, in opposition to Western science, which was characterised by rationalism and materialism, Ibáñez defended the existence of a 'true' science which, in being a 'total, full science' could only exist in close proximity to God (Ibáñez, 1940, pp. 17–18).

The second fundamental characteristic of this Spanish science that aspired to God was its emphasis on the unity of different sciences (Santana, 2009 and 2013). Indeed, the CSIC's founding act called for 'the restoration of the classic, Christian unity of the sciences that was destroyed in the eighteenth century' (BOE, 28 November 1939, p. 6668), thus re-establishing the balance between speculative and experimental science. This unity of science to which the CSIC aspired was by no means an attempt to reconcile the two cultures in the sense that C.P. Snow (1959) advocated, and had nothing whatsoever to do with the neopositivist wish for a unified science. In reality, the aim was neither to reconcile nor rebalance, but rather to hierarchise (a term the National-Catholics used profusely), since the hierarchisation of values was a key part of their philosophy. For National-Catholicism, modern societies were characterised by an erroneous subversion of values, in which material values had displaced spiritual and transcendental ones from their former position of power and governance. At an individual level, the medieval *whole man*, characterised by an appropriate hierarchy of values, had broken apart following the moral and cultural fracturing of the Renaissance, giving rise to incomplete human beings (Pemartín, 1942, pp. 71–2). In the scientific field, this subversion was manifested through the emancipation of knowledge regarding the material world from its ages-old subordination to religion. In the words of the Minister, 'once philosophy, the unifying force in the world of science, had been dethroned from its monarchical seat, the discipline of the physical and biological was destroyed in respect to the spiritual, as if matter and life were not themselves governed by the spirit, and as if our knowledge of things had not to be unified by natural reason in the highest causes' (Ibáñez, 1940, p. 19). What followed, then, was a re-hierarchisation that subjugated materialist science to the dictates of speculative sciences, understood from a religious perspective. As late as 1947, the Minister began his book on Spanish research by underscoring the hierarchisation embodied by the CSIC and the pre-eminence at its core of those sciences which focused on the domains of the spirit (Ibáñez, 1947, p. 9).

This pre-eminence in the post-war CSIC of the science of the spirit, as conceived from the perspective of religion, has been overshadowed by the adjective *scientific* which was added to the name of the new institution. In 1930s Spain (and even now in common Spanish), the term *science* referred to a form of empiricist, materialist knowledge based on reason and experimentation, as opposed to speculation, ideology or religious beliefs. Consequently, the name of the previous institution, the JAE, made a distinction between the generic term *studies* and the much more specific term *scientific research*. The founders of the CSIC, however, opted to intentionally manipulate the usual semantic field of the term in order to introduce into the name of the new organisation an anachronistic conception of science, understood as knowledge in general in which religion played a key guiding role. Indeed, this new meaning that the National-Catholics aimed to bestow on the term *science* was so forced and unnatural that they were eventually obliged to add further adjectives (Spanish, imperial, full, total, true, and so on) in order to clarify what exactly they meant by this new concept and to distinguish it from what was commonly understood by science, at the risk of rendering their discourse unintelligible. This twisting of concepts highlights the novel nature of the CSIC, in the sense that it broke away from the science projects considered previously by the victors.

In his reflections, Lora Tamayo used the term *science* in its most common and usual sense. The expression 'Spanish science' simply meant the science conducted in Spain; it was never used as a cognate for a science of a different nature or substance. Lora's concern was more closely linked to the relationship between pure and applied science. Even the Institute of Spain proposed by Minister Sáinz did not attempt to alter the traditional meaning of *science*. Quite the contrary, in fact, since he did not identify science with knowledge or culture, the Minister was forced to find a specific task for science within the structure of this organisation, assigning the centres of the now dissolved JAE to the office of a vice-presidency created *ad hoc* for Palacios. In this sense, the structure of the Institute implicitly presupposed a distinction between culture, including the scientific Royal Academies that needed to be reoriented and reshaped in order to adapt them to the New Spain, and scientific research centres, which by contrast would continue to function in accordance with their own criteria once the ideological loyalties of their scientists had been ensured following the purge. Palacios himself assumed the existence of neutral, objective scientific criteria when he opposed the purging of the Spanish Society for Physics and Chemistry. The implicit view seemed to be that scientists must be purged since they were employees of the State, but that science itself was neutral and objective, or at least was located on a fairly autonomous plane in relation to ideology and politics (Gómez and Canales, 2009, p. 326).

This was precisely the kind of implicit outlook to which Ibáñez and Albareda aimed to put a stop. For National-Catholicism, it was unacceptable to suppose that a form of knowledge could exist outside the value-based and transcendental domain. It was therefore necessary to strip the concept of science of its scientific connotations and call all forms of knowledge *science*, despite the confusion that this semantic manipulation generated. Only by twisting meanings in this sense was it possible to proceed with the hierarchisation of knowledge. In short, the inclusion of the adjective 'scientific' in the name of the new council was no mere coincidence, but rather a highly revealing indication of the concept of science that the authorities were striving to impose.

Spanish Science and the Structure of the CSIC

The unity of science, conceived in theological terms, was not simply a rhetorical resource used to satisfy the Catholic Church and the fundamentalist or traditionalist factions; it was one of the guiding principles of the CSIC. So much so, in fact, that the structure of the new organisation was based precisely on this concept of science. The structure was inspired by Ramon Llull's tree of science, which is nurtured by sacred science, and which became the CSIC's logo. In the words of the Minister, 'the imperial tree of science grew and flourished in the imperial garden of catholicity and sacred and divine science sat comfortably on its throne, as its essential fibre and nerve, providing the juice that nourished all the branches in unison' (Ibáñez, 1940, p. 17).

Religion presided over the whole institutional framework of the CSIC, so much so, in fact, that the Minister based the structure of the new council on the work of the Supreme Creator. Each of the three areas of this work (matter, life and spirit) were assigned two big centres within the institution (CSIC, 1942, p. 43). The study of the physical world (matter) was placed under the control of the Alfonso el Sabio Centre, which was mainly theoretical in nature, and the Juan de la Cierva Centre, which was more focused on applied science. Biological studies (life) were the responsibility of the Santiago Ramón y Cajal and Alonso de Herrera Centres. And finally, to oversee studies in the fields of humanities and social science (spirit), the Raimundo Lulio and Marcelino Menéndez Pelayo Centres were established. Each of these centres in turn encompassed several research institutes. Following the National-Catholic logic of the unity of science, the Institute of Theology was established as a kind of umbrella body. Thus, an all-encompassing structure was achieved which included all areas of knowledge.

Studies focusing on the CSIC have tended to pay little attention to this totalising structure, since it is hard to explain in accordance with interpretative suppositions upon which such works are based, as explained in the introduction. The proliferation of institutes related to theology, biblical studies and humanities does not seem to fit in with a CSIC designed to serve as an instrument for fostering autarkic science. Indeed, the institutes which were related in one way or another to this perspective, including the pure sciences and economics, hardly accounted for half of the total initial structure. This circumstance also fails to mesh with the interpretation that stresses the continuity of the new council in relation to the JAE, since the all-encompassing nature of the CSIC had little in common with the structures of the former JAE and FNICER. The structure of the JAE was not a global, systematic and hierarchical design, but rather the circumstantial addition of centres and institutions that existed at any given moment. Thus, for example, despite the central position afforded to education in its activities (Students' Residence, the experimental high school, grants for international study visits related to educational themes, and so on) and the influence of the members of the ILE, the JAE had no centre or institute specifically dedicated to pedagogic thought, since the National Pedagogic Museum was an independent body. The CSIC, on the other hand, had its own Institute of Pedagogy, created *ex novo* to control and hierarchise the competences of these dispersed, and now dissolved, organisations. The initial structure of the CSIC does not, therefore, seem to have been either a continuation of the JAE or an attempt to respond to the needs of the autarky. Rather, it was a reflection of the unity of science from a religious perspective, and more specifically, an attempt to hierarchise all knowledge under the guiding light of religion.

The Control of the Scientific Communities

This all-encompassing and systematic structure based on the unity of science turned the CSIC into a powerful apparatus for controlling knowledge in the New Spain and for containing and categorising the scientific and academic communities that had survived the process of purge and exile. A number of different factors indicate that this was indeed the main objective of the CSIC. In this sense, one thing that is particularly striking at the outset is the sharp contrast which existed between the Council's large-scale institutional development and its low level of research activities. During the initial years, the institutes had only a few grant students and the directorships were mainly filled by professors who continued lecturing at their respective universities. Work in the institutes was only carried out in the afternoons (Santesmases, 1998, p. 315). Research remained a non-professional pursuit up until 1945, when the figure of scientific collaborator was established, followed by that of senior researcher in 1947.

Nevertheless, this lack of research and researchers proved no obstacle to the profuse institutional development of the CSIC. In 1945 there were more than forty research institutes, almost twice as many as the 23 that were in existence in 1940, despite the fact that this number itself was already very high (Canales, 2009, p. 122). This proliferation of research institutes with no or little accompanying research suggests that the CSIC institutes were not conceived primarily as research units, but rather as control hubs for redesigning their respective scientific and academic communities. Research in a specific field was far less important than the creation of an institute directed by trusted scientists who would ensure proper control of any potential future development.

This preventative control strategy was also deployed in other key areas of the institutionalisation of the scientific communities, such as the publication of journals. Despite the dire economic situation, the CSIC continued to publish a large number of journals; some were dependent on the JAE, others were reassigned to the CSIC and a few, such as the *Revista Española de Pedagogía* (Spanish Pedagogic Journal), were clear attempts at replacing independent journals now no longer up and running. In 1940, the CSIC published 22 journals, of which only seven had existed before the war; by 1942 this figure had almost doubled (42) and by 1945 the number of journals published had risen to an astounding 55 (Malet, 2008, p. 242). By publishing a vast number of journals in a wide range of disciplines, the CSIC made a significant contribution to the development of different fields, but more importantly, it also helped control the main backbone of their respective communities.

Power of decision over who was authorised to represent Spain at international conferences helped strengthen this control over scientific and academic communities, a control which was already fairly strong due to discretionary powers regarding grants for studying abroad. The CSIC even established its own awards, tellingly called the Francisco Franco awards, through which it acknowledged scientific merit in the New Spain.

This notable institutional development was accompanied by the building of a new campus over the main pre-war facilities inherited by the CSIC. It seems that for the victors, dismantling the structure of the JAE was not enough; they wanted to eliminate all physical traces of its very existence by constructing new, monumental buildings that would attest to the grandeur and magnificence of the new organisation (Canales, 2012). Thus, the CSIC became even more of an empty shell. As Santiago López points out, there were more buildings than resources for research, more container than content (López, 1997, p. 205).

In short, even before any real research was initiated in a specific field, the CSIC had created the instruments for its future control; an institute, a journal, and even a building, all under the directorship of trusted scientists appointed in accordance with their ideological or religious beliefs (Canales, 2009, pp. 128–35), who monopolised all relations with international communities by controlling who was authorised to participate in international conferences and by hand-picking candidates for study abroad grants. This desire to gain preventative control over research makes sense of the major push for institutional development devoid of content which was so characteristic of the CSIC in the immediate post-war period, and provides answers to the questions that other interpretations are at a loss to explain. The aim was not to conduct research, but rather to control the playing field in order to prevent others from doing so.

Faced with the immense control apparatus designed by the National-Catholics, the traditional freedom of scientists vanished into thin air. During his speech at the opening ceremony of the CSIC, the Minister clearly stated that from a Thomistic view of 'science for truth and goodness' there was no room for 'intellectual efforts on the wrong path', since 'errors cannot be allowed to constitute science, and no scientific liberty can exist for them' (Ibáñez, 1940, p. 17). Nevertheless, the Minister also acknowledged that universities and scientific activity themselves did require some degree of autonomy. He therefore proposed a compromise based on the subordination of 'free scientific initiative' to the vigilance of the State, which would be responsible for safeguarding political unity and preventing scientific activities from 'being used, under any circumstances, as a perverse instrument against the sacred principles of our Homeland' (Ibáñez, 1940, p. 33). Scientists would therefore have technical autonomy in their research, but both the establishment of general research areas and any political dimension to their work would be the direct responsibility of the State.

Nevertheless, this authoritarian compromise proved particularly unstable at the theoretical level since, as a result of the National-Catholic axiom, the nation identified itself with religion and since, in its view, religion aspired to a kind of alternative science, even this restricted technical autonomy was called into question. The solution proposed by the Minister to the issue of scientific freedom was again a twisting of meanings, in this case in relation to the concept of freedom. Freedom lost its traditional meaning, and was now understood as an acknowledgement of the truth as established by the victors. As the Minister stated in his inaugural speech of 1940: 'There is no more academic freedom or freedom of teaching, except that of the truth of imperial catholic Spain, the only truth that makes us free ...' (Ibáñez, 1940, p. 33). In 1942 he distorted the semantic field even more by recalling the time of the Empire as a period in which 'freedom flourished at the service of Spain and Science, which is a truer and higher freedom than the rigid, narrow channelling of scientific activities at the behest of a handful of researchers with too much freedom' (Ibáñez, 1942, p. 102). His predecessor at the ministry, Pedro Sáinz Rodríguez, had earlier expressed much the same idea, albeit somewhat less rhetorically: 'we are not enemies of freedom; we are, however, enemies of the liberal interpretation of freedom because we believe that the liberal interpretation of freedom is not true human freedom' (quoted by López Bausela, 2011, pp. 122–3). It seems obvious, then, that the National-Catholics entertained their own concept of freedom, which was a far cry from the traditional meaning of the term. Whatever the case, what the Minister was trying to make crystal clear to scientists was that 'gone are the days in which one's status as an intellectual could be used as an excuse for offences against the Homeland, as a justification for violating the basic pillars of the State ...' (Ibáñez, 1940, p. 17). Obviously, in practice, the technical autonomy of the new scientific elite did in fact survive, although it should not be forgotten that this elite had been hand-picked by the Regime for their ideological and religious sympathies.

The Failure of the Utopia: the Pragmatic Commitment to the Material World

Despite the formidable control apparatus that had been designed, the National-Catholic project, embodied in the CSIC, encompassed a contradiction that was difficult to reconcile: the tacit acceptance of science's contribution to the material development of the country. In truth, this issue was something the authorities preferred to gloss over if possible. The material dimension of the products of science took up very little space in National-Catholic discourses, relegated to second place by the overriding importance of imperial, spiritual and religious concerns. It is extremely significant that the founding act of the council only mentions science's contribution to national development in fourth place, after three openly ideological objectives linked to an alternative science: the submission of science to Spain's historic evolution, the training of new scholars in accordance with this aim and the aspiration to contribute to universal culture. And even in fourth position, when scientific research was placed at the service of national interests, the text refers to 'spiritual and material' interests, with spiritual being mentioned first (BOE, 28 November 1939, p. 6668). The material dimension of science therefore occupied only a very minor place in the discourse of those responsible for Spain's new science policy.

Nevertheless, the configuration of the new regime made it difficult to ignore this material dimension completely. For a start, the regenerationist discourses which had talked so much about Spain's lack of scientific development as the main cause of the country's backwardness had not fallen on deaf ears, even among the more conservative sectors. Also, in a regime established as the result of a military victory and led by soldiers, the military uses of science was an issue that continued to arouse a fair degree of interest. It is no coincidence that the origins of the regenerationist discourses were rooted in the colonial defeat of 1898. Finally, the falangists did not share the spiritual priority defended by the National-Catholics. As good fascists, they aspired to a material rather than spiritual Empire, and to the material aggrandisement of the nation.

In his inaugural speech, the Minister was forced to climb down from his lofty perch of imperial rhetoric and acknowledge that 'it is not enough to ensure the rebirth of the universal value of imperial science' (Ibáñez, 1940, p. 21); it was also necessary to ensure a science that contributed to the material progress of the nation. The solution was a merging of spirit and matter, in order to enable 'a science defined in accordance with the canons of our golden culture, but at the same time set within the demands of the present and obediently subject to the yoke of national interest and prosperity' (Ibáñez, 1940, p. 16). Nevertheless, not even Ibáñez's characteristic rhetoric could reconcile the contradiction underlying this approach: it was obvious to all that the science that would bring about this material prosperity was not Spanish imperial science, but rather science understood in the common, traditional sense. And this science was none other than the reviled science of Encyclopaedism, which had generated the crisis of 'our science' and had attacked the 'unshakable principles of our Homeland' (Ibáñez, 1939, p. 2); it was the science of the 'heretical scientists who caused the rivers of our national genius to become parched and dry, and who plunged us into lethargy and decline' (Ibáñez, 1940, p. 17); in short, it was the much-slandered 'false European scientism' (Ibáñez, 1940a, p. 7). Thus, this acknowledgement of the material dimension of science rendered the National-Catholic discourse absurd, since it entrusted the progress of the nation to the same forces that had, in its view, brought about its decline.

This tension between National-Catholics, with their defence of spiritualisation, and the advocates of a science that would contribute to the material progress of the nation was ongoing, and can still be detected in 1942. In the inaugural speech of that year, the Minister attacked all those who sought to recover values instead of acknowledging the progress made since the war; in other words, he attacked those who 'mourn for this and that which has been lost, rather than congratulating themselves for so much gain' (Ibáñez, 1942, p. 102).

At that time still, a simple reference to the 'role of cold-hearted intellectuality in the sowing of discordance' was enough to deflect any veiled criticism of the spiritualising National-Catholic project. However, the position of this sector was strengthened by the Regime's autarkic economic policy and by the international scenario which was increasingly hostile to the fascism with which the Regime identified itself.

This evolution can be clearly perceived in the inaugural speech of the academic year given at the University of Valencia in 1944. On this occasion, the Minister reformulated his proposal of 1939, in which he had advocated synthesis, introducing here a number of significant changes: 'The scientific heights attained by Spain during the Golden Age should be our spur; but this should, in turn, be subject to the demands of the current moment in our country, and to the imperial interests of our century' (Ibáñez, 1944, p. 17). Imperial science was, therefore, no longer the aim, but rather simply a spur – a stimulus subject moreover to the demands of the country in the present moment. It was a painful concession which sat uncomfortably with the Minister, which is no doubt why he then went on to highlight the fact that the new aim did not affect the supremacy of religion, and insisted that the priority was still:

> above all, to spiritualize the task of the researcher or scientist at all times. Because although it is true that we want to view Spanish Science as a factor that contributes to the material progress of the country, we state, above all, that our scientific movement will come to nothing if we do not also conceive of it as an effort of intelligence to ensure the supreme possession of the Truth. (Ibáñez, 1944, p. 17)

Despite these rhetorical exercises, in 1944 the Minister himself was forced to acknowledge the failure of the reactionary utopia which underpinned the creation of the CSIC, and to reach a pragmatic compromise that would safeguard the original programme as much as possible. The new strategy abandoned the priority of fostering an alternative, religion-based science and opted instead for a defensive strategy aimed at preventing science, understood in its traditional sense, from questioning the ideological values of National-Catholicism. Instead of striving to promote Spanish imperial science, the aim now was simply to control the ideology of scientists and their activities.

This radical change from an ambitious policy aimed at establishing an alternative science to a much more prosaic policy of defence was a clear indication of the total failure of the reactionary utopia. Nevertheless, this failure in no way called the CSIC itself (or its initial structure) into question. Quite the opposite, in fact, the control mechanisms established as part of the initial project became particularly useful in guaranteeing a scientific community which, despite working on the material dimension of science, never for one moment questioned the ultimate aim of the scientific activity in which they were engaged, which continued to be 'the supreme discovery of God' (Ibáñez, 1944, p. 18). The CSIC became firmly established as an enormous safety net responsible for putting a preventative stop to any scientific development that posed a potential threat to the values of National-Catholicism and for conferring a transcendent dimension on materialist scientific practice. As the Minister himself declared: 'having achieved this transcendent dimension for our Science, there is no risk now in applying its practice to the physical world and the dimensions of matter and life' (Ibáñez, 1944, p. 18).

Conclusion: Technocracy *Avant la Lettre*

The National-Catholic authorities did not succeed in establishing their reactionary utopia of a restored National-Catholic science. Like all utopias, it was simply unattainable. Nevertheless, this failure does not justify overlooking the attempt itself when interpreting the CSIC, for a number of reasons. Firstly, because if this utopia, which guided the whole process, is not taken into account, it is impossible to explain the creation and configuration of the CSIC. Secondly, and importantly, because the configuration of the CSIC that arose from the National-Catholic project of scientific unity survived, despite the failure of the project itself. And finally, because the abandonment of the National-Catholic project did not call into question the control mechanisms established to watch over Spain's various scientific communities and their activities. On the contrary, after the failure, these mechanisms became even more important, as a means of guaranteeing a science devoid of any connotations that potentially went against religion, the dictatorship or the reigning social order.

The result was that Spanish science was one step ahead of other areas in the search for a way out of the contradiction that, from 1945 onwards, threatened the survival of a regime based on the principles of the European dictatorships from the inter-war period in an international scenario which had been erected precisely on the foundations of their defeat. The solution in the scientific field was a commitment to technified science, a science which was without doubt materialist and rationalist, but was stripped of all the anti-religious or emancipating connotations that many associated with the science of the pre-war period. The onset of the Cold War facilitated this technocratic conception of science. The agreements signed with the USA in 1953 freed the Regime from international ostracism and enabled access to technology and the re-entry of Spanish scientific communities into the international arena. It was in this framework that some exiled scientists were able to return to their native country, providing they agreed to limit themselves to the technical questions posed by their work and to ensure careful avoidance of any ideological implications. Paradoxically, the reviled notion of a neutral, objective science that did not transcend the material aspect of its focus of study ended up being the chosen solution of the Franco Regime, in the same way as the *apolitical* government of technocrats, centred on the economic development, ended up being the chosen solution of the Regime itself just one decade later.

List of References

Camprubí, L., 2014. *Engineers and the Making of the Francoist Regime.* Cambridge MA: The MIT Press.

Canales Serrano, A.F., 2009. 'La política científica de posguerra'. In: A. Gómez Rodríguez and A.F. Canales Serrano (eds), 2009. *Ciencia y fascismo. La ciencia española de posguerra.* Barcelona: Laertes, pp. 105–36.

Canales Serrano, A.F., 2012. 'A new space for a new science: the transformation of the JAE Campus after the Spanish civil war. *History of Education*, 41(5), pp. 657–74.

Claret, J., 2006. *El atroz desmoche.* Barcelona: Crítica.

CSIC, 1942. *Memoria de la Secretaría General, 1940–1941.* Madrid: CSIC.

García Camarero, Er. and García Camarero, En. (eds), 1970. *La polémica de la ciencia española.* Madrid; Alianza.

Giral, F., 1994. *Ciencia española en el exilio (1939–1989). El exilio de los científicos españoles.* Barcelona: Anthropos – Centro de Estudios Republicanos.

Gómez Rodríguez, A., 2014. 'Frontera e integridad en el contrato social para la ciencia española, 1907–1939'. *Dymanis*, 34(2), pp. 415–87.

Gómez Rodríguez, A. and Balmer, B., 2013. 'Ciencia y política: una cuestión de frontera'. In: A. Gómez Rodríguez and A.F. Canales Serrano (eds), 2013. *Estudios políticos de la ciencia. Políticas y desarrollo científico en el siglo XX.* Madrid: Plaza y Valdés. pp. 15–34.

Gómez Rodríguez, A. and Canales Serrano, A.F., 2009. 'The rebels and the new Spanish scientific culture'. *Journal of War and Cultural Studies*, 2(3), pp. 321–33.

Gutiérrez Ríos, E., 1990. 'El Consejo Superior de Investigaciones Científicas. Su gestación y su influjo en el desarrollo científico español'. *Arbor*, 529, pp. 77–99.

Ibáñez Martín, José, 1939. *La Universidad actual ante la cultura hispánica. Discurso pronunciado en 23 de octubre de 1939 – año de la victoria – en el paraninfo de la Universidad Central, en la apertura del curso académico 1939–1940*. Madrid: Prensas de Silverio Aguirre.

Ibáñez Martín, José, 1940. 'Discurso pronunciado con ocasión de la sesión inaugural del Consejo Superior de Investigaciones Científicas'. In: J. Ibáñez Martín, 1947. *La investigación española. Tomo I*. Madrid: Publicaciones Españolas. pp. 11–38.

Ibáñez Martín, José, 1940a. *Discurso pronunciado por el Excelentísimo Sr. Ministro de Educación Nacional D. José Ibáñez Martín en el acto académico solemne, conmemorativo del XXV aniversario de la fundación de la universidad*. Murcia: Sucesores de Nogués.

Ibáñez Martín, José 1940b. *Hacia un nuevo orden universitario. Discurso pronunciado por el Excmo. Sr. D. José Ibáñez Martín, Ministro de Educación Nacional, en la inauguración del curso académico de 1940–41 en la Universidad de Valladolid*. Valladolid: Establecimiento Tipográfico de Samarán.

Ibáñez Martín, José, 1942. 'Discurso pronunciado con ocasión del Tercer Pleno del Consejo Superior de Investigaciones Científicas'. In: J. Ibáñez Martín, 1947. *La investigación española. Tomo I*. Madrid: Publicaciones Españolas. pp. 75–105.

Ibáñez Martín, José, 1944. *Renacimiento científico en al investigación y en la docencia. Discurso pronunciado por el Excmo. Sr., José Ibáñez Martín, ministro de Educación Nacional, en la solemne inauguración de la Facultad de Ciencias de la Universidad de Valencia, el día 7 de octubre de 1944*. Valencia: Universidad de Valencia.

Ibáñez Martín, José, 1947. *La investigación española. Tomo I*. Madrid: Publicaciones Españolas.

López Bausela, J.R., 2011. *La contrarrevolución pedagógica en el franquismo de guerra. El proyecto político de Pedro Sainz Rodríguez*. Madrid: Biblioteca Nueva.

López García, S., 1997. 'El Patronato Juan de la Cierva (1939–1960): I Parte: Las Instituciones Precedentes'. *Arbor*, 157(619), pp. 201–38.

López García, S., 1998. 'El Patronato Juan de la Cierva (1939–1960): II Parte: La organización y la financiación'. *Arbor*, 159(625), pp. 1–44.

López García, S., 2008. 'Las ciencias aplicadas y las técnicas: la Fundación Nacional de Investigaciones Científicas y Ensayos de Reformas y el Patronato

Juan de la Cierva del CSIC'. In: A. Romero de Pablos and M.J. Santesmases (eds), 2008. *Cien años de política científica en España*, Madrid, Fundación BBVA. pp. 79–106.

Lora Tamayo, M., 1939. *Investigación dirigida. Ideas sobre una ordenación racional de la investigación científico-técnica*. Sevilla: Imprenta municipal.

Maiocchi, R., 1993 'Scientiati italiani e scienza nazionale (1919–1939)'. In: S. Soldani and Turi, G. (eds), 1993. *Fare gli italiani. Scuola e cultura nell'Italia contemporanea. II. Una società di massa*. Bolonia: Il Mulino, pp. 41–86.

Malet, A., 2008. 'Las primeras décadas del CSIC: investigación y ciencia para el franquismo'. In: A. Romero de Pablos and M.J. Santesmases (eds), 2008. *Cien años de política científica en España*. Madrid: Fundación BBVA. pp. 211–56.

Malet, A., 2009. 'José María Albareda (1902–1966) and the formation of the Spanish Consejo Superior de Investigaciones Científicas'. *Annals of Science*, 66(3), pp. 307–32.

Mann, M., 2004. *Fascists*. New York: Cambridge University Press.

Morente, F., 2005. 'La universidad fascista y la universidad franquista en perspectiva comparada'. *Cuadernos del Instituto Antonio de Nebrija*, 8, pp. 190–96.

Morente, F., 2015. 'Entre tinieblas. La universidad española en la larga posguerra'. In: A.F. Canales Serrano and A. Gómez Rodríguez (eds), 2015. *La larga noche de la educación española. El sistema educativo español en la posguerra*. Madrid: Biblioteca Nueva. pp. 183–217.

Otero Carvajal, L.E., dir. 2006. *La destrucción de la ciencia en España*. Madrid: Editorial Complutense.

Pemartín Sanjuán, J., 1937. *Qué es 'lo nuevo'... Consideraciones sobre el momento español presente*. Sevilla: Tip. Alvarez y Zambrano.

Pemartín Sanjuán, J., 1942. 'Significado y alcance de la Reforma de la Enseñanza Media'. In: J. Pemartín Sanjuán, 1942. *Formación clásica y formación romántica. Ideas sobre la enseñanza*. Madrid: Espasa-Calpe.

Puig-Samper Mulero, M.A. (ed.), 2007. *Tiempos de investigación. JAE-CSIC, cien años de ciencia en España*. Madrid: CSIC.

Santana de la Cruz, M., 2009. 'Unidad de la patria, unidad de la ciencia. La retórica científica del régimen franquista'. In: A. Gómez Rodríguez and A.F. Canales Serrano (eds), 2009. *Ciencia y fascismo. La ciencia española de posguerra*. Barcelona: Laertes. pp. 165–86.

Santana de la Cruz, M., 2013. 'Las políticas de la ciencia en España: de la República a la Dictadura'. In: A. Gómez Rodríguez and A.F. Canales Serrano (eds), 2013. *Estudios políticos de la ciencia. Políticas y desarrollo científico en el siglo XX*. Madrid: Plaza y Valdés. pp. 35–52.

Santesmases, M.J., 1998. 'El legado de Cajal frente a Albareda'. *Arbor*, 160(631–2), pp. 305–32.

Snow, C.P., 1959. *The Two Cultures and the Scientific Revolution*. New York: Cambridge University Press.

Suñer, E., 1937. *Los intelectuales y la tragedia española*. Burgos: Editorial Española.

Serial Reference

BOE – *Boletín Oficial del Estado*

Chapter 5

Broken Science, Scientists under Suspicion. Neuroscience in Spain during the Early Years of the Franco Dictatorship[1]

Rafael Huertas

Following the Spanish Civil War (1936–39), the Franco Regime initiated a wide-ranging ideological offensive aimed at rejecting democratic values and imposing an authoritarian system influencing every aspect of political life and Spanish society. Meanwhile, a campaign was launched to discredit and suppress the institutions most closely identified with the Second Republic (1931–39). The creation on 24 November 1939 of the Higher Council for Scientific Research (*Consejo Superior de Investigaciones Científicas* – CSIC) should be understood in this context as an interruption of the tradition represented by the Board for Advanced Studies and Scientific Research (*Junta para Ampliación de Estudios e Investigaciones Científicas* – JAE), in an attempt to reorient the areas and disciplines of scientific research in the direction determined by Nationalist and Catholic thinking (Santana de la Cruz, 2013).

The new scientific and social paradigm that emerged after the Civil War gave rise to complex situations that allow us to assess Spanish scientific activity in a context of political repression, with the removal of 'Republican' scientists and the reorientation of scientific policies and institutions (Otero, 2006). The social context of hunger, illness and poverty among broad swathes of the population also led to serious public health problems that had to be addressed (Arco, 2006). A case of special interest is that of the neurosciences, both basic and applied. The aim of this article is to analyse the situation of the neurosciences in Spain during the early years of the Franco Regime – the 1940s – by studying two distinct but complementary cases. Firstly, the process of 'political cleansing' and the refounding of the Cajal Institute, whose prestige gave it legitimacy on

---

[1] This research was made possible thanks to Research Project HAR2012–37754–C02–01 (Ministry of Economy and Competitiveness – Spain).

the international stage and had to be taken into account by the authorities. Secondly, the development of epidemiological, clinical and laboratory research into neurological illnesses relating to vitamin and nutritional deficiencies and directly connected to the post-war context of hunger and shortages.

The Purging of the Cajal Institute

The Biological Research Laboratory, founded in 1901 and directed from the outset by Santiago Ramón y Cajal (González de Pablo, 1998), was the forerunner to what would later become the Cajal Institute (Baratas and Santesmases, 2002, pp. 68–9; Baratas, 2007). Ramón y Cajal – who won a Nobel Prize in 1906 – played a decisive role in Spanish scientific activity over the first third of the twentieth century in at least three aspects. On a purely scientific level he headed up a leading neurohistological school, which expanded into neurohistopathology, neurophysiology and so on, on the basis of his neuron theory (López-Muñoz, Boya and Álamo, 2006). At an institutional level, as president of the JAE (1907), Cajal gave his seal of approval to a number of science policy and researcher training initiatives that were begun and enhanced during the Second Republic (López-Ocón, 2007). Finally, on a symbolic level, Cajal – who carefully positioned himself outside specific party affiliations and interests (Lewy, 1987, p. 98), transcended the frontiers of his own specialism to embody the virtues of the ideal scientist, unselfish wise man, and undisputed master. Cajal's moral authority boosted the perception of science as a flag bearer of progress and an essential tool for the modernisation of the country (Otero and López Sánchez, 2012, pp. 193ff.), and for the renewal of the Fatherland. However, this connection between Science and Fatherland – so evident in Cajal – would among many of his disciples become Science and the Republic, or even Science and Revolution. Faced with the Fascists' irrational cry of 'death to the intelligence' the identification of the Republic with science and culture, as a beacon of reason, was a constant refrain in the discourse of the Republican sector during the Civil War (Salmón and Huertas, in press).

Thus, the Cajal Institute was without a doubt one of Spain's most prestigious scientific institutions, though one that seemed to be closely tied to the liberal, secular culture represented by the JAE, or even to the 'anti-Spanish', materialist and 'Leftist' spirit of the Republic itself, according to the victors of the Civil War, which placed the liberals and Leftists in the same camp. A good example of such rhetoric is provided by declarations Francisco Franco (1892–1975) himself made to the Havas news agency on 27 August 1938, a few months before victory was declared:

> The war in Spain is not an artificial affair: it is the crowning moment of a historical process, the struggle of the fatherland against those who would deny it, of unity against backwardness, of morality against crime, of the spirit against materialism, and there is no other solution than the triumph of these pure and eternal principles over bastardized, anti-Spanish beliefs. (Franco, 1975, vol. I, p. 50)

Those in command of the new state believed it necessary to conduct a 'political cleansing' that would purge the Institute of its 'undesirable' connotations but without renouncing the benefits of the international prestige Cajal and his school had garnered. Thus, at the newly-created Higher Council for Scientific Research (CSIC), research in the fields of medicine and animal biology was carried out in dependent institutes of the Santiago Ramón y Cajal Centre, including most notably the Cajal Institute itself and the National Institute for Medical Sciences, created in 1942 (Huertas, 2007). As may be seen, the name of Ramón y Cajal continued to provide a nominal lustre to these institutions, but such gestures barely concealed the total dismantling of his school. The strategy pursued for this end may be summed up by two specific actions. Firstly, Cajal's most outstanding disciples were retired from any position of responsibility. Secondly, a change was imposed on the structure of the Institute meaning it no longer dealt exclusively with the neurosciences but was opened up to other areas that had little or nothing to do with its original objectives.

As for the scientific staff of the Institute, some left the country to work and die in exile. Thus, Dionisio Nieto (1908–85) settled in Mexico, where he first worked at the *La Castañeda* asylum before becoming head of the Psychiatric and Brain Research service of the National Institute for Neurology and Neurosurgery (Sacristán, 2007). Miguel Prados Such (1899–1970) joined the Neurological Department at McGill University in Montreal, where he was Professor of Psychiatry, founded the Montreal Psychoanalytical Club in 1946 and was very active in the foundation of the Canadian Psychoanalytical Society (Parkin, 1987). Finally, Gonzalo Rodríguez Lafora (1886–1971) was in Mexico – where he taught conferences, worked as an advisor and ran a prestigious private consultancy – for a brief exile from which he returned in 1947 (Valenciano, 1977, p. 161). The purpose here is not, of course, to provide a blow-by-blow account of the fates of all the scientists exiled from the Cajal Institute, but rather to demonstrate the impact they came to have upon the academic milieu of their receiving countries (Fernández Guardiola, 1997; Díaz, 2009).

However, other of Cajal's disciples remained in Franco's Spain and were submitted to purging processes, with varying outcomes. Francisco Tello (1880–1958), who had replaced Cajal at the head of the Institute, was retired from this role as well as from the Chair in Histology and Pathological Anatomy at the Central University of Madrid, which he had held since 1926. In the purging process he was accused of being an atheist, of not having lost his position during the war, of having signed the intellectuals' manifesto against the national army following the bombardment of Madrid, of having occupied leading posts such as Dean of the Faculty of Medicine, and of not having cooperated in the 'triumph of the Glorious Uprising'.[2] Tello's punishment finally arrived with an ambiguous formula: 'change of activity for a similar one', which in practice prevented him from teaching. He was allowed, however, to continue his work at the CSIC, as long as he held no position of responsibility. Only in 1949–50, when he was practically on the point of retirement, did he have his chair restored (González Santander, 2005, p. 50).

Francisco Tello, Cajal's favoured disciple, is a good example of what has been called 'internal exile'. Internal exile may be defined as the situation experienced by a large group of doctors and scientists who remained in Spain after the Civil War without maintaining political or ideological links or direct contact with the Franco Regime. It is a diverse group that ranges from those who suffered more or less violent forms of repression, to those viewed with indifference by the Regime, and who were not harassed, but nor did they receive any support in the pursuit of their professional careers (Giral, 1994, pp. 48ff.).

As may be imagined, many variations exist within this general paradigm of the 'internal exile'. Upon his return from Mexico, Rodríguez Lafora had to face the same purging process, in which he was accused of being a Leftist, being anticlerical and of belonging to the Association of Liberal Doctors (Moya, 1986, pp. 121ff.). In 1944, before his return to Spain, he was prohibited from practising medicine for eight years and told to pay a fine.[3] For this reason, upon his return in 1947 he was unable to return to his position as head of the psychiatric service of Madrid's Hospital Provincial, though he did rejoin the Cajal Institute, where he became head of the Pathological Anatomy of the Nervous System department, created in 1951 (CSIC, 1952, p. 185).

Finally, the fate of another of Cajal's leading disciples is of interest. Fernando de Castro (1896–1967) received a far more benevolent treatment than many

---

[2]   Archivo General de la Administración (AGA) Education section, 31/4001. The 'Charge Sheet' that sets out these accusations is dated 8 November 1939.

[3]   AGA, Justice section, 42/30513.

of his colleagues. His purging process file concludes 'no penalty'[4] and he was able to hold on to his Chair in Histology at the University of Seville, until he transferred to Madrid in 1951. He continued to pursue his research work at the Cajal Institute, although he was denied the position of Director. After the war, he was the representative of the Cajal school of neurohistology who held the most international standing and prestige. In 1948 he travelled to a number of countries in South America (Argentina, Chile, Uruguay and Peru) delivering lectures and classes in academic institutions (CSIC, 1950, pp. 165–6).

Once the disciples of Cajal had been removed from positions of responsibility at the Institute, its running was taken over by scientists who enjoyed the trust of Franco's Regime, even if their specialisms had nothing to do with the neurosciences. The first of these was the paediatrician Enrique Suñer (1878–1941). Suñer, advisor on Public Education during the dictatorship of Primo de Rivera (1923–30), made a name for himself as an ideologist of Franco's new state with a book entitled *Los intelectuales y la tragedia española* (The Intellectuals and the Spanish Tragedy) (1937), which fiercely criticised the Free Institution for Teaching and the JAE. He had also been vice-president of the Culture and Teaching Commission of the State Technical Council (governing body of the insurgent camp during the Civil War). Upon the conclusion of hostilities he was restored to his position as Chair of Paediatrics at the University of Madrid, and accumulated a number of positions, including President of the General Assembly of the Spanish Red Cross and of the General Council of Medical Colleges, and Director of the National School of Childcare, as well as heading up the Cajal Institute, as mentioned. However, Suñer died in 1941, before he could get to grips with all of these positions. Assigned to replace him at the head of the Cajal Institute was Juan Marcilla (1886–1950), Chair of Viticulture and Oenology since 1924, and of Agricultural Microbiology since 1928, at the School of Agronomical Engineering of Madrid.

During the Second Republic, Marcilla had been designated Director of the Wine Research Centre, part of the National Foundation of Scientific Research and Trials of Reform (*Fundación Nacional para Investigaciones Científicas y Ensayos de Reformas* – FNICER), which aimed to link up basic research with the business sector. José Castillejo (1877–1945), Director of the FNICER and secretary of the JAE, instigated the creation in 1933 of the Wine Research Centre with the aim of furthering research into a sector of such importance to the Spanish economy (Formentín and Rodríguez, 2001). This apparent accommodation with the Republican system of science earned him a

---

4   AGA, Education section, 31/20312.

'no penalty' purging process, and he was able to return to his university chair without any trouble, himself arbitrating on purging processes at the School of Agronomical Engineers in August 1939, the war having ended on April 1 the same year (Santesmases, 2001, p. 47). When the Cajal Institute underwent restructuring in 1940, Marcilla took charge of the newly created Department of Fermentation, before becoming its head a year later, as stated. Other departments were created at the same time, including Chemistry, Biology, Physiology and Organic Reactions (CSIC, 1942, p. 201). These scientific disciplines were not related to the neurosciences pursued by Cajal's school.

Throughout this process a fundamental role appears to have been played by the general secretary of the CSIC, José María Albareda (1902–66). Albareda held a doctorate in pharmacy, a specialism in soil science, and had been a secondary school teacher. At the end of the war he secured the Chair in Applied Mineralogy at the University of Madrid's Faculty of Pharmacy and was named secretary-general of the CSIC, a position he held for life and which he used to shape and direct the scientific policy of the institution (León-Sanz, 2009; Malet, 2009). He was a member of the *Opus Dei* and was ordained as a priest in 1959, although he never gave up his scientific work. His position as a 'scientific Catholic' is still vaunted by bodies close to the *Opus Dei* (Pérez-López, 2010).

On the one hand, as we have seen, Albareda placed trusted scientists – Catholics who were loyal to the Regime – at the head of the Cajal Institute, which to some extent acted as a counterbalance to the scientific and symbolic import of the neurohistopathology being studied there. However, it should be borne in mind that Albareda's own interest in supporting and developing scientific fields close to him as a pharmacist and soil scientist meant he could compete with the histologists (familiar with microbiology), whom he did his best to exclude (Santesmases, 1998). Marcilla was the one to leave the Cajal Institute, to become the director of the new Institute of Applied Microbiology that was created in 1946, the year the Society of Spanish Microbiologists was founded (Carrascosa, 2008). Thus, together with the political reasons mentioned, we cannot disregard corporate elements that favoured the development of particular scientific disciplines and the creation of research institutes connected to the areas of interest of the secretary-general, and always run by trusted cronies.

Despite the lack of interest by the Franco Regime's scientific policy in the basic neurosciences, the heirs to the Cajal school continued to work in the Institute during the 1940s and 1950s. Thus for example Francisco Tello pursued embryological research into the formation of the cranial and spinal nerves and on the development of the sympathetic nervous system (Tello, 1946, 1947, 1949), while Fernando de Castro worked on the peripheral nervous system (reflex arcs, visceral innervation, synaptic transmission, and so on) (De Castro, 1942, 1950).

The Department of Organic Reactions directed by Julián Sanz Ibáñez (1909–63), which soon became the Virus Department, deserves special mention. Sanz Ibáñez was the only scientist to have belonged to the old Cajal Institute to receive support from the new administration. He was secretary of the Institute from 1941 until 1946, the year in which he became its director, following the aforementioned transfer by Marcilla. He combined these activities with his teaching as Chair of Pathological Anatomy at the Faculty of Medicine in Madrid. His notable research includes that on exanthematic typhus – isolated as various strains and which acted as a contact point for the technicians from the Rockefeller Foundation in Spain (Rodríguez Ocaña, 2001) – but above all his work on poliomyelitis, both in the microbiological context (isolation and study of the biological properties of the polio virus) (Ballester and Porras, 2009) and in the histological context (histopathology of the alterations in the peripheral nervous system, experimental poliomyelitis, and so on) (Sanz Ibáñez, 1945). Sanz Ibáñez became a reference point in Spain for virology research into poliomyelitis, and maintained and worked on the SK New Haven strain in the Instituto Cajal between 1941 and 1951. This strain was later given to the Virology Laboratory at the National Hygiene Institute as a result of the epidemic outbreaks in the 1950s (Albadalejo, 1958).

Without question, as we have attempted to show, the early years of the Franco dictatorship saw the dismantling and 'refounding' of the Cajal Institute. This involved a strategy that not only placed individuals loyal to the Regime at the head of the institution but also diluted and undermined the singular character of its research activities with the introduction of new scientific disciplines. The subject of neurohistology was treated as to be extinguished. It is only since the 1970s that a renewed interest in the basic neurosciences has emerged, though now based on other foundations such as biochemistry or molecular biology (Santesmases, 2001, p. 68).

The Neurology of Hunger

While the basic or 'laboratory' neurosciences were undergoing the institutional changes described above, clinical neurology had to confront a major challenge: the notable rise in deficiency neuropathies and other neurological illnesses due to the lack of food during the first two years of the Franco dictatorship (Del Cura and Huertas, 2007). The connection between nutrition and neurological illness had become clear during the Civil War, notably through the studies carried out in Republican Madrid during the *no pasarán* (they shall not pass) period by the nutritionist Francisco Grande Covián (1909–95), the neurologist Manuel Peraita (1908–50) and the psychiatrist Bartolomé Llopis (1905–66) from 1937 at the National Institute for Food Hygiene, which had been created at the start of the war. These studies combined an epidemiological focus with dietary data and direct observation of patients at the Polyclinic for deficiency disorders that was established in this Institute (Del Cura and Huertas, 2008).

However, the results of this research were published after the end of hostilities in a very different political context. These were, as we have just noted, scientific studies hosted by the Republican government – something that was systematically ignored – and whose authors were submitted to purging processes. Franco's new state used a number of strategies to appropriate this research.

On the first pages of the *La alimentación en Madrid durante la guerra* (Feeding in Madrid during the war), published by Grande Covián a few months after the end of the war in December 1939, its author explains his reasons for carrying out the work in the following terms:

> Having long held an interest in the question of nutrition and finding myself forced to remain in Madrid during the war, my original intention was to study the dietary regime to which we were subject, and its consequences for the health of Madrid's residents. (...) Our purpose in presenting these studies is not only scientific. We also want to make known with all due objectivity the extreme lack of humanity of the authorities of Red Spain, who were fully aware of the lack of food in Madrid. (Grande Covián, 1939, pp. 1–2)

It does not seem that Grande Covián's concern with declaring himself a 'sympathizer' of the Regime has much to do with a political or ideological acceptance of the victors, but rather with an imperative to distance himself from the Republican authorities. Grande Covián had to face a purging process that imputed to him a series of charges, such as having served the government of the Popular Front in a number of positions of responsibility (all scientific or academic), as secretary of the Medicine Faculty or as part of the assistant management of the above mentioned Food Hygiene Institute. Moreover, and given his knowledge of several languages, he was accused of acting as an interpreter before foreign commissions that sought to discredit the 'National Uprising'. The sentence handed down by the Law of Political Responsibilities in May 1940 prohibited him from occupying management and confidential positions and he was prevented from standing for professorships, obtaining bursaries and study grants or from occupying positions connected to teaching for four years. In July 1940 Grande Covián requested that his file be reviewed with resulting 'exculpatory' claims, which was rejected (Pérez Peña, 2005, p. 258).

In October of the same year, Carlos Jiménez Díaz (1898–1967), Chair of Medical Pathology at the University of Madrid and one of the most prestigious doctors under Franco (Peset, 2000), offered him a position at the Faculty of Medicine's Institute of Clinical and Medical Research (attached to his Chair), which had been created in April that year, thereby joining the Santiago Ramón y Cajal Centre of the Higher Council for Scientific Research (Jiménez Díaz, 1965). During the early 1940s, under the orders of Jiménez Díaz, Grande Covián was the contact point for technical staff from the International Health Division of the Rockefeller Foundation in the development of major nutrition surveys carried out in working-class districts of Madrid (Del Cura and Huertas, 2009). In 1950 he finally obtained the Chair of Physiology at the University of Zaragoza, though shortly thereafter he transferred to the Laboratory of Physiological Hygiene at the University of Minnesota, in the city of Minneapolis, where he was to remain for 20 years (Gómez-Santos, 1992, pp. 161ff.). During his period in the United States he worked with and became a member of the editorial board of the journal *Ciencia*, the voice of Spanish scientists in exile (Puig-Samper, 2001; Aleixandre, Micó and Soler, 2003).

To return to the nutritional and clinical studies carried out during the siege of Madrid, in the middle of the Civil War Grande Covián co-authored together with Manuel Peraita a monograph entitled *Avitaminosis y Sistema Nervioso* (Avitaminosis and the Nervous System) which represents the most significant contribution made to the deficiency neuropathies of the Civil War. Most of the neurological disorders observed were due to a deficiency of vitamins in the B2 group, which are normally obtained through foods of animal origin (meat, eggs, milk), which were almost wholly absent from the diet of the residents of Madrid after 1937. At the time, numerous cases of pellagra began to manifest themselves together with a series of neurological and psychiatric symptoms associated with this illness (Grande Covián and Peraita, 1941, p. 39).

Bartolomé Llopis was without doubt the scholar who contributed most to the study of neuropsychiatric disorders related to pellagra during the war and post-war period, and he did so with dedication and care. Designated head doctor of the women's Psychiatric Clinic at Madrid's Provincial Hospital during the war, he possessed significant clinical data that was supplemented by patients from the Polyclinic for deficiency disorders, enabling him to investigate in depth the mental disorders connected to pellagra.

Llopis was a genuine representative of what we earlier referred to as 'internal exile'. A militant of the Republican Left since 1934, and medical captain of the Republican Army, he had to face a purging process in 1940 that prohibited him from practising medicine for four years. In 1944 he was allowed to return to professional practice and in 1946 he occupied, on an interim basis, a position as head of clinic – on the orders of López Ibor (1908–91) – at the Psychiatric service of Madrid's Provincial Hospital, taking charge of the men's section until 1950, when he handed over responsibility to Gonzalo Rodríguez Lafora, whom he had always considered his mentor and who was rejoining the hospital following his return from Mexico. It is of great interest to note that between 1940 and 1944, when Llopis was earning his living working in a telegraph office, he was able to publish three scientific articles on pellagra (Llopis, 1940, 1941, 1943), indicating not only his own interest in keeping up his academic activity and working on the material gathered during the war, but also the intellectual and professional recognition of his colleagues, 'sympathizers of the regime', who enabled his articles to be published in the journals *Actas Españolas de Neurología y Psiquiatría* (founded and edited by López Ibor) and in the *Revista Clínica Española* (founded and edited by Jiménez Díaz).

It was not until 1946, however, that he published *La psicosis pelagrosa. Un análisis estructural de los trastornos psíquicos* (Pellagra psychosis. A structural analysis of mental disorders), which attempts to present a broad synthesis of his work and theories combined with data from clinical observation. In his introduction Llopis does not engage in an 'act of conscience', or avail of the occasion to reject the 'red barbarism' in an attempt to curry favour with the new state. On the contrary, he merely states that:

> The hunger suffered by residents of Madrid during the last Spanish Civil War gave rise to an 'epidemic' of pellagra of an extent and intensity perhaps never seen before (...) I considered it my duty to avail of this unfortunate situation to investigate the specific characteristics of Pellagra-related mental disorders (Llopis, 1946, p. 11).

It is worth emphasising that the model of pellagra served Llopis to propose, in a wide-ranging and influential text published in 1954 in *Archivos de Neurobiología*, a psychopathological theory in which he reviews and updates the concept of unitary psychosis (Llopis, 1954a, 1954b).

Thus, as we have indicated, the two monographs *Avitaminosis y Sistema Nervioso* (1941) and *La psicosis pelagrosa* (1946) brought together the results of an investigation carried out in the 'red zone', with the support of the Republican government and in a situation of hunger and harassment suffered by those who would subsequently lose the war. These circumstances were disregarded by the victors and used to publish the studies as the product of the scientific excellence of the new state. It is no coincidence, in this regard, that Juan José López Ibor, one of the Regime's most prestigious and powerful psychiatrists, was to write the prologues to both monographs. It was a way of endorsing the authors, but also of symbolically appropriating their work.

A final, but significant, contribution to neurology resulting from the studies carried out in Madrid during the war was made by Manuel Peraita with his description of clinical systems he termed 'paresthesia-causalgia syndrome' or 'Madrid syndrome', whose key symptom was an acute – and extremely painful – sensation of burning on the soles of the feet, and which was related, as with pellagra, to vitamin deficiency. Although the syndrome was described for the first time in the abovementioned monograph, *Avitaminosis y Sistema Nervioso* (1941), Peraita published a brief note on his research in the *British Medical Journal* (1946) – which led to his work winning worldwide recognition – and an extended version in Spanish a year later (Peraita, 1947). The experience of the Spanish Civil War offered keys for interpreting neurological pathologies in situations of war and/or hunger. It was soon related to the 'burning feet syndrome' described by Gopalan (1946) in poor regions in southern India, and by Simpson (1946) as affecting British prisoners in the Far East. Other scholars alluded specifically to Peraita's contribution in work on neuropathies among prisoners of war (Smith, 1946), while John D. Spillane, in his *Nutritional Disorders of the Nervous System* (1947), highlighted the importance of the Spanish experience in the study of deficiency neuropathies, mentioning the 'Paresthesia-Causalgia Syndrome' and adding that: 'These terms were used by Spanish workers to designate conditions in which the prominent features consisted of spontaneous dysaesthesiae of the extremities and trunk' (Spillane, 1947, p. 136). The international significance of the work carried out by Manuel Peraita into this important neurological pathology is thus established (Huertas and Del Cura, 2010).

Unlike his colleagues in the siege of Madrid, there were no consequences for Peraita as a result of his purging process. As his file explains, during the war Peraita helped persecuted individuals of a religious background and of 'nationalist interest'. He held no positions of responsibility in Republican institutions, nor did he collaborate with the authorities save in a strictly professional sense. With regard to his work on deficiency neuropathies, he declared that he had not handed over the results of his research, but saved them for publication in nationalist Spain.[5] Regardless of the more or less truthful rhetoric he may have employed in his statements, the significance of the fact that these studies ended up being published in Franco's Spain is interesting.

After the Civil War, Peraita worked at the Neurophysiological Laboratory of the Cajal Institute of Biological Research. In 1943, with the support of López Ibor, he was designated Director of the Leganés state asylum (close to Madrid), a position he held until 1949 (Conseglieri, Villasante and Del Cura, 2007, p. 558). He died at an early age in 1950.

Another neurological disease related to nutrition that gave rise to interesting academic work in the 1940s was neurolathyrism. Lathyrism is a chronic illness resulting from the excessive and continuous consumption of vetch (*Lathyrus sativus*) and that causes paralysis of the lower body. Between 1941 and 1943, an epidemic of neurolathyrism occurred as a direct result of the socioeconomic conditions prevailing during the first few years of the Franco dictatorship, which was a period of political repression but also of hunger and deprivation for large swathes of the population. Epidemiological studies made it possible to establish that the disease principally affected young men who worked as day labourers or industrial workers and whose diet consisted almost solely of vetch. The disease did not affect the wealthier classes, whose diet was more varied, or the very poor – and malnourished – who could not afford the staple. However the most interesting point is that the public health problem forced the authorities to carry out a series of clinical and laboratory studies that contributed to a better understanding of this neurological disease (Del Cura and Huertas, 2009a). The group led by Carlos Jiménez Díaz once again took on the burden of the research and became a benchmark for doctors in the rest of the country (Del Cura, 2004).

A final aspect of the neurosciences under Franco that deserves mention relates to the neurosurgeon Sixto Obrador (1910–79). During the war years and in the immediate post-war period he was not in Spain, but in England pursuing his studies (with a grant provided by JAE since 1934). When the Second World War broke out he left for Mexico, where he completed his training as a neurosurgeon and contributed to the journal *Ciencia*. Giral (1994, p. 257) includes him on his list of exiled doctors, though his personal circumstances are not those of the typical exiled Republican. He returned to Spain in 1946 and, under the auspices of Jiménez Díaz, founded the Madrid Institute of Neurosurgery and went on to create and organise the neurosurgery services that became established in the country's leading hospitals (Gutiérrez and Izquierdo, 1998).

---

[5]   AGA Education section, 31/2213.

Conclusion

As has been shown in the preceding pages, after the Spanish Civil War there was a reorientation of scientific policies which, in the case of the neurosciences, followed well-defined strategies: in the first place, the purging and exile – both internal and external – of the leading scientists of the Republican period and their replacement in management positions by Franco loyalists who were not necessarily specialists in the nervous system. In the field of basic neurosciences an attempt was made to break with the tradition – secular, liberal and Republican – represented by the Cajal school, without thereby losing the prestige and resulting legitimacy granted by his Nobel Prize. Secondly, a degree of 'appropriation' took place by the 'science of the new state' of the results of significant research carried out during the Civil War but only published in the early years of the dictatorship. Finally, some of these scientists were 'accommodated' by leading Regime figures in medicine and were able to carry on working beneath their 'protection', in a competent manner but in very difficult conditions, with little recognition and always 'under suspicion'.

List of References

Albadalejo, L., 1958. 'Investigaciones sobre el virus de la poliomielitis'. *Medicamenta*, 8, pp. 3–7.

Aleixandre, R., Micó, J.A. and Soler, A., 2003. 'La contribución científica del exilio a través de la revista *Ciencia (1940–1975)*'. In: J.L. Barona (ed.), 2003. *Ciencia, salud pública y exilio (España 1875–1939)*. Valencia: Seminarid'Estudis sobre la Ciencia, pp. 73–98.

Arco, M.A. del., 2006. '"Morir de hambre". Autarquía, escasez y enfermedad en la España del primer franquismo'. *Pasado y Memoria. Revista de Historia Contemporánea*, 5, pp. 241–58.

Ballester, R. and Porras, M.I., 2009. 'El significado histórico de las encuestas de seroprevalencia como tecnología de laboratorio aplicada a las campañas de inmunización. El caso de la poliomielitis en España'. *Asclepio*, 61(1), pp. 55–80.

Baratas, A., 2007. 'Neurociencias en la Junta para Ampliación de Estudios'. *Asclepio*, 59(2), pp. 115–36.

Baratas, A. and Santesmases, M.J., 2002. *Nobeles españoles. De la neurona al ADN*. Madrid: Nívola.

Carrascosa, A., 2008. 'Juan Marcilla. Presidente fundador de la SEM'. *Actualidad SEM*, 45, pp. 16–21.

Conseglieri, A., Villasante, O., Del Cura, M.I., 2007. 'El manicomio nacional de Leganés en la posguerra. Aspectos organizativos y clínico-asistenciales'. In: R. Campos, L. Montiel and R. Huertas (eds), 2007. *Medicina, ideología e historia en España (siglos XVI–XXI)*. Madrid: CSIC, pp. 555–68.

Consejo Superior de Investigaciones Científicas, 1942. *Memoria de la Secretaría General, 1940–1941*. Madrid: Consejo Superior de Investigaciones Científicas.

Consejo Superior de Investigaciones Científicas, 1950. *Memoria 1948*. Madrid, Consejo Superior de Investigaciones Científicas.

Consejo Superior de Investigaciones Científicas, 1952. *Memoria 1951*. Madrid: Consejo Superior de Investigaciones Científicas.

De Castro F., 1942. 'Modelación de un arco reflejo en el simpático, uniéndolo con la raíz aferente central del vago. Nuevas ideas sobre la sinapsis'. *Trabajos del Instituto Cajal de investigaciones biológicas*, 34, pp. 217–301.

De Castro F., 1950. 'Die normale Histologie des peripheren vegetativen Nervensystems. Das Synapsen-Problem: Anatomisch-experimentelle Untersuchungen'. *Verhandlungen der Deutschen Gesellschaft für Pathologie*, 34, pp. 1–52.

Del Cura, M.I., 2004. *La epidemia de latirismo en España. Aspectos epidemiológicos, clínicos y sociales*. PhD. Universidad Autónoma de Madrid.

Del Cura, M.I. and Huertas, R., 2007. *Alimentación y enfermedad en tiempos de hambre. España, 1937–1947*. Madrid: CSIC.

Del Cura, M.I. and Huertas, R., 2008. 'The siege of Madrid (1937–1939). Nutricional and clinical studies during the Spanish civil war'. *Food and History*, 6(1), pp. 193–214.

Del Cura, M.I. and Huertas, R., 2009. 'Public Health and Nutrition after the Spanish Civil War. An Intervention by the Rockefeller Foundation'. *American Journal of Public Health*, 99(10), pp. 1772–9.

Del Cura, M.I. and Huertas. R., 2009a. 'Describiendo el neurolatirismo. Los clínicos ante la epidemia de latirismo en la España de la posguerra'. *Revista de Neurología*, 48(5), pp. 265–70.

Díaz, J.L., 2009. 'El legado de Cajal en México'. *Revista de Neurología*, 48, pp. 207–15.

Fernández Guardiola, A., 1997. *Las neurociencias en el exilio español en México*. Mexico DF: Fondo de Cultura Económica.

Formentín, J. and Rodríguez, E., 2001. *La Fundación Nacional para Investigaciones Científicas (1931–1939)*. Madrid: CSIC.

Franco, F., 1975. *Pensamiento político de Franco: antología.* Sistematización de textos por Agustín del Río Cisneros. Madrid: Ediciones del Movimiento.

Giral, F., 1994. *Ciencia española en el exilio (1939–1989): El exilio de los científicos españoles.* Barcelona: Anthropos.

Gómez-Santos, M., 1992. *Francisco Grande Covián. El arte y la ciencia de la nutrición.* Madrid: Temas de Hoy.

González de Pablo, A., 1998. 'El Noventa y ocho y las nuevas instituciones científicas. La creación del laboratorio de Investigaciones Biológicas de Ramón y Cajal'. *Dynamis*, 18, pp. 51–79.

González Santander, R., 2005. *La escuela histológica española VII.* Madrid: Cersa.

Gopalan, G., 1946. 'The burning feet syndrome'. *Indian Medical Gazette* (Calcutta), 81, pp. 22–6.

Grande Covián F., 1939. *La alimentación en Madrid durante la Guerra. (Estudio de la dieta suministrada a la población civil madrileña durante diecinueve meses de guerra: Agosto 1937 a febrero de 1939).* Madrid: Publicaciones de la Revista de Sanidad e Higiene Pública.

Grande Covián, F. and Peraita, M., 1941. *Avitaminosis y Sistema Nervioso.* Madrid: Ed. Miguel Servet.

Gutiérrez, D. and Izquierdo, J.M., 1998. *El doctor Obrador en la medicina de su tiempo.* Oviedo: Bear.

Huertas, R., 2007. 'Las ciencias bio-médicas en el CSIC durante el franquismo'. In: M. Puig-Samper (ed.), 2007. *Tiempos de investigación JAE-CSIC, cien años de ciencia en España.* Madrid: CSIC, pp. 293–7.

Huertas, R. and Del Cura, M.I., 2010. 'Deficiency Neuropathy in Wartime: The "Paraesthetic-Causalgic Syndrome" Described By Manuel Peraita During the Spanish Civil War'. *Journal of the History of the Neurosciences*, 19, pp. 173–81.

Jiménez Díaz, C., 1965. *Historia de mi Instituto.* Madrid: Paz Montalvo.

León-Sanz, M.P., 2009. 'Science, State and Society: José María Albareda's Consideraciones sobre la investigación científica'. *Prose Studies*, 31(3), pp. 227–40.

Lewy, E., 1987. *Santiago Ramón y Cajal. El hombre, el sabio y el pensador.* Madrid: CSIC.

Llopis, B., 1940. 'La psicosis pelagrosa, la psicopatología general y la nosología psiquiátrica'. *Actas Españolas de Neurología y Psiquiatría*, 1(3–4), pp. 174–93.

Llopis, B., 1941. 'Algunas consideraciones generales sobre la pelagra'. *Revista Clínica Española*, 3(4), pp. 328–34.

Llopis, B., 1943. 'Los trastornos psíquicos de la pelagra (Enfermedad de Casal)'. *Actas Españolas de Neurología y Psiquiatría*, 4(1–2), pp. 7–31.

Llopis, B., 1946. *La psicosis pelagrosa. Un análisis estructural de los trastornos psíquicos*. Barcelona-Madrid-Valencia: Científico Médica.

Llopis, B., 1954a. 'La psicosis única'. *Archivos de Neurobiología*, 17(1), pp. 3–41.

Llopis, B., 1954b. 'La psicosis única (conclusión)'. *Archivos de Neurobiología*, 17(2), pp. 141–63.

López-Muñoz, F., Boya. J. and Álamo, C., 2006. 'Neuron theory, the cornerstone of neuroscience, on the centenary of the Nobel Prize award to Santiago Ramón y Cajal'. *Brain Research Bulletin*, 70(4–6), pp. 391–405.

López-Ocón, L., 2007. 'La voluntad pedagógica de Cajal, presidente de la JAE'. *Asclepio*, 56(2), pp. 11–36.

Malet, A., 2009, 'José María Albareda (1902–1966) and the formation of the Spanish Consejo Superior de Investigaciones Científicas'. *Annals of Science*, 66(3), pp. 307–32.

Moya, G., 1986. *Gonzalo R. Lafora. Medicina y cultura en una España en crisis*. Madrid: Universidad Autónoma de Madrid.

Otero Carvajal, L.E. (ed.), 2006. *La destrucción de la ciencia en España. Depuración universitaria en el franquismo*. Madrid: Editorial Complutense.

Otero Carvajal, L.E. and López Sánchez, J.M., 2012. *La lucha por la modernidad. Las ciencias naturales y la Junta de Ampliación de Estudios*. Madrid: CSIC-Residencia de Estudiantes.

Parkin, A., 1987. *History of psychoanalysis in Canada*. Toronto: Toronto Psychoanalytical Society.

Peraita, M., 1946. 'Deficiency Neuropathies Observed in Madrid During the Civil War (1936–1939)'. *The British Medical Journal*, 23, p. 784.

Peraita, M., 1947. 'El complejo sintomático de Madrid: Síndrome parestésico-causálgico'. *Revista Clínica Española*, 26, pp. 225–40.

Pérez-López, P., 2010. 'José María Albareda: La ciencia al servicio de Dios'. *Nuestro tiempo*, 665, pp. 52–7.

Pérez Peña, F., 2005. *Exilio y Depuración (en la Facultad de Medicina de San Carlos)*. Madrid: Vision Net.

Peset, J.L., 2000. 'Carlos Jiménez Díaz, maestro de la medicina española'. *Eidon*, 5, pp. 13–15.

Puig-Samper. M.A., 2001. 'La revista *Ciencia* y las primeras actividades de los científicos españoles en el exilio'. In: A. Sánchez and S. Figueroa (eds), 2001. *De Madrid a México. El exilio español y su impacto sobre el pensamiento, la ciencia y el sistema educativo mexicano*. Morelia: Universidad Michoacana de San Nicolás de Hidalgo/Comunidad de Madrid, pp. 95–125.

Rodríguez Ocaña, E., 2001. 'La reorientación de la intervención de la Fundación Rockefeller en Salud Pública en España (1931–1941)'. In: *XXIst International Congress of History of Science*. Mexico, 8–14 July 2001.

Sacristán, C., 2007. 'En defensa de un paradigma científico. El doble exilio de Dionisio Nieto en México, 1940–1985'. In: R. Campos, Villasante, O. and R. Huertas (eds), 2007. *De la "Edad de Plata" al exilio. Construcción y "reconstrucción" de la psiquiatría española*. Madrid: Frenia, pp. 327–46.

Salmón, F. and Huertas, R., in press. 'Unifying science against Fascism: Neuropsychiatry and medical education in the Spanish Civil War (1936–1939)'. In: H. Kamminga and G. Somsen (eds). *Pursuing the unity of science: Scientific practice, ideology between the Great War and the Cold War*. Aldershot: Ashgate Publishing.

Santana de la Cruz, M., 2013. 'Las políticas de la ciencia en España: de la República a la Dictadura'. In: A. Gómez and A. Canales (eds), 2013. *Estudios políticos de la ciencia. Políticas y desarrollo científico en el siglo XX*. Madrid: Plaza y Valdés, pp. 35–51.

Santesmases, M.J., 1998. 'El legado de Cajal frente a Albareda: las ciencias biológicas en los primeros años del CSIC y los orígenes del CIB'. *Arbor*, 160(631–2), pp. 305–32.

Santesmases, M.J., 2001. *Entre Cajal y Ochoa. Ciencias biomédicas en la España de Franco, 1939–1975*. Madrid: CSIC.

Sanz Ibáñez, J., 1945. 'Estudio experimental de la poliomielitis'. *Revista de Sanidad e Higiene Pública*, 20(3), pp. 215–30.

Simpson, J., 1946. 'Burning feet in British prisoners of war in the Far East'. *Lancet*, 1, pp. 959–61.

Smith, D., 1946. 'Nutritional Neuropathies in the Civilian Internment Camp Hong Kong, January, 1942–August, 1945'. *Brain*, 69, pp. 209–22.

Spillane, J.D., 1947. *Nutritional Disorders of the Nervous System*. Edinburgh: Livingstone.

Suñer, E., 1937. *Los intelectuales y la tragedia española*. San Sebastián: Editorial Española.

Tello, F., 1946. 'Sobre la formación de los ganglios nerviosos craneales y el mesectodermo, en el embrión de pollo'. *Trabajos del Instituto Cajal de investigaciones biológicas*, 38, pp. 1–40.

Tello, F., 1947. 'La evolución de la cresta neuronal y su relación con los ganglios espinales en el embrión de pollo'. *Trabajos del Instituto Cajal de investigaciones biológicas*, 39, pp. 1–79.

Tello, F., 1949. 'Lo evidente y lo dudoso en la génesis del simpático, con nuevas observaciones'. *Trabajos del Instituto Cajal de investigaciones biológicas*, 41, pp. 1–107.

Valenciano, L., 1977. *El Doctor Lafora y su época*. Madrid: Morata.

Chapter 6

Cultures of Research and the International Relations of Physics Through Francoism: Spain at CERN[1]

Xavier Roqué

This chapter analyses Spain's first stage at the European Organization for Scientific Research (CERN), the transnational high-energy research facility established by 12 European countries in 1954. Spain became a member of CERN on 1 January 1961, withdrew in June 1969 with effect from 31 December 1968, and joined the organisation again in 1983, after the country's transition to democracy. It is the only member state to have left and re-entered the organisation. The episode itself has been little researched. The three-volume *History of CERN* reflects the nearly negligible impact of Spain's membership in the 1960s: Spain first appears in vol. 2 as a country 'hardly represented at Geneva in the mid-1960s', and is listed in vol. 3 as one of the countries that abstained on the new convention in 1967 (Pestre, 1990a, p. 409; Krige, 1996, p. 70).[2] Spanish high-energy physicists have read the episode in political and economic terms, as 'a paradigmatic example of erratic governmental policy', while placing themselves aside from the decision (Aguilar and Ynduráin, 2003, p. 19; Aguilar, 2004, 2007). Spain would have joined CERN mostly for diplomatic reasons and would have left it as its contribution became unbearable. Even though 'membership clearly had a positive scientific impact', it was 'badly planned and nothing was done to take

[1] This work was supported by the Spanish Ministry of Economy and Competitiveness through research project 'Física, cultura y política en España' (HAR2011–27308), and is based on Roqué (2012). Unless otherwise stated, all archival references are to the CERN Archives, Geneva (Switzerland). I acknowledge permission to quote from the Royal Society Archives and the CERN archives, and thank CERN archivist Anita Hollier for her assistance.
[2] The index of vol. 2 does not include the references to Spain in table 7.11, 'Staff in grade 10 or higher in 1962 by division and country' (Pestre, 1990a, p. 403) and table 11.3, 'The distribution by country of the value of contracts awarded by CERN in the years 1952–1965 inclusive' (Krige, 1990, p. 668).

advantage of it' (Pascual, 1998, pp. 234–5). Historian Albert Presas grants that Spain's first stage at CERN 'is difficult to understand, to say the least', and also that withdrawal reflected 'the lack of a science policy able to capitalize on the benefits of membership' (Presas, 2007, p. 303).

One important work has implicitly challenged the physicists' account. Building on the archives of the Spanish Ministry of Foreign Affairs, Ana Romero and José Manuel Sánchez Ron (2001, pp. 203–6) have provided evidence of the Spanish government's commitment to the organisation through the 1960s. Central to their account is the Spanish candidature for the site of the new Super Proton-Synchrotron that CERN started to plan in 1964. In June 1966, a commission including chairmen of industry, high-rank officials, diplomats, scientists and an embedded journalist, campaigned in CERN for the Spanish site, a visit that was well publicised in the Spanish press. However, in spite of the unity exhibited in Geneva, Franco's cabinet was eventually split on the candidature. In September 1967 a dedicated commission advised the government only to remain in CERN if the new accelerator was built in Spain; for the Ministry of Foreign Affairs, membership was not to be conditioned by the decision about the site; and the Ministry of Finance pushed for withdrawal whatever the case. Romero and Sánchez Ron finally endorse the view that the Spanish economy could not afford the increasing quota, so that leaving the organisation was inevitable (2001, pp. 203–26).

I would like to add new layers of research and meaning to this episode and to make it speak to the themes of this book. Building on recent research on physics through Francoism, as well as on untapped CERN archives and personal archives of some of the physicists involved, I will argue that the development and outcome of Spain's first stage at CERN was not just dictated by the diplomatic and economic possibilities of Franco's Regime, but also by the research culture and practices built after the Civil War (1936–39). Conversely, this episode may help us probe into this culture and practices. Beyond the mounting quota and diplomatic convenience, what can be said of the experience of Spanish physicists at CERN? To what extent were the physicists willing and able to adopt the investigative ethos of a big scientific installation? Historians and sociologists of science have shown that the practice of contemporary physics requires enculturation (Traweek, 1992; Pestre, 1984; Low, 2005; Kaiser, 2005) and scientists themselves have on occasion acknowledged and discussed it too. An important document in the CERN archives, dated 1963 and dealing with the cooperation between CERN and its member states, refers to the need for scientists to adopt the 'scientific culture' and 'habits of mind' required by high-energy physics research:

A physicist, however brilliant and however well learned and competent he may be, is to some extent lost when he first comes to CERN if he has not had previous experience in this field ... The habits of mind created by having to fit into a machine time-table, or to accept that any modification of apparatus may be a solid engineering job, is, for people who are going to play a leading part in research at CERN, perhaps as important and as difficult to acquire as experience in the particular techniques and modes of thought of high-energy work.[3]

We need to take these modes of thought and research cultures into account in order to understand Spain's first stage at CERN and the international relations of Spanish science through Francoism more generally.

The chapter is in three parts. I first discuss the relations between physics and culture in Franco's Spain, which I will argue are essential to make sense of Spain at CERN. I then show how these cultural relations bore upon two prominent issues that had to be balanced against each other: the contribution of member states, on the one hand, and the recruitment of staff and the tender of contracts, on the other. I will argue that Spanish diplomats, together with scientists qua policy makers such as José María Otero Navascués, did not lack political will, but dealt with CERN in autarkic terms, on the basis of national quotas. Ultimately, their institutional and in Otero's case personal strategy, rooted in the autarkic policies of early Francoism, did not pay off with CERN the way it had with other international bodies.

Physics and Culture in Franco's Spain

In the wake of the Spanish Civil War (1936–39), physicists in Spain took advantage of, and were used by General Francisco Franco's Regime, to promote international relations and further the autarkic economy. Herran and Roqué (2013) have described three important ways in which physics and the Regime helped each other through the first two decades of Francoism: moulding the community, aligning the discipline and seizing the institutions. First, the Regime's policy makers used war, exile and purging to mould the physicists' community around military-related and applied fields such as optics and the material and nuclear sciences. The vacancies left by the exiles provided ample professional opportunities for physicists showing the right political allegiances.

---

[3] 'Co-operation between CERN and member states', CERN/482/Rev. 2, 27 May 1963 (DIR-ADM-FIN-06 (3) FILE 1, p. 2).

Second, the key physicists for the reconstruction of the discipline after the Civil War shared a pronounced outlook on the cultural value of contemporary physics, which can be traced back to the pre-war Republican period. Ideologues on the extreme right, together with conservative physicists and philosophers, sought to replace the progressive modernist reading of physics that had arguably prevailed in Spain before the Civil War, with a reactionary modernist reading that stressed the spiritual rather than the material dimensions of the discipline and argued for its integration into the Christian scheme of the world. Third, both the cultural realignment of the discipline and the adaptation of physicists to the new intellectual environment were of a piece with the institutional changes that harnessed the physical sciences to the military and economic needs of the autarkic state. This process resulted in a form of physics that was most obviously autarkic in its relation to the economy and power, but also autarkic because national norms of excellence and cultures of research prevailed over international ones.[4]

Our interpretation of the rise of physics during the dictatorship eschews both wholesale condemnation and uncritical recognition, seeking rather to locate the discipline and its practitioners within the Francoist political economy and political order, along with recent work on science and Francoism (Camprubí, 2014; Claret, 2006; Gómez and Canales, 2009; Malet, 2008 and 2009; Romero and Santesmases, 2008). Engineers such as J.M. Otero Navascués (1907–83) and physicists such as Carlos Sánchez del Río (1924–2013), who both played an important role in Spain's first stage at CERN, are representative of the relations I have described. Indeed, contrary to the physicists' account, Spain's membership entailed high political stakes. It was mediated by the Spanish Nuclear Energy Board (*Junta de Energía Nuclear* – JEN), a governmental agency created in 1951 and presided by Franco's head of general staff and Minister of Air Force, Juan Vigón Suerodíaz (1880–1955). Vigón was succeeded by Navy engineer J.M. Otero Navascués, a key figure in the reconfiguration of physics in the decades following the Civil War, as the Regime sought to enlist modern science in the construction of a totalitarian state based on a National-Catholic tradition. Crucially, in the case of the physical sciences, this did not entail the rejection of contemporary theories such as relativity or quantum mechanics, but rather their spiritualisation. Thus, by revealing the limits of human understanding,

---

[4]   I build here on the meanings of autarky as discussed by Saraiva and Wise (2010) in their introduction to an issue of *Historical Studies in the Natural Sciences* on fascism and science, comparing the experience of Germany, Italy, Portugal, France and Spain. On the limits of autarky and the necessity of international exchange, see Guirao (1998). See also Anduaga (2009) and Santesmases (1999).

indeterminism and acausality justified religious guidance and eased the integration of physics into the "scheme of creation" (Otero Navascués, quoted in Herran and Roqué, 2012, p. 220).

Costs: Spain's Contribution to CERN

There is no adequate historical account of the process leading to Spain's membership of CERN. The Spanish Nuclear Energy Board (JEN) led an early attempt to join the organisation in 1954, but the CERN Council decided 'not to admit new members until the first machines were in operation'. On the wake of the Geneva Conference on the Peaceful Uses of Atomic Energy in 1955, Otero regretted 'the failure of international cooperation at CERN', insisting on the need to join the organisation as soon as the restriction was lifted.[5] Having successfully staged the celebration of the 10th anniversary of the CSIC in 1950 with ample international representation, and having signed the Concordat with the Vatican and the Pact of Madrid with the United States in 1953, the Regime felt confident it could restore international relations without renouncing its reactionary values. Foreign scientists were elected as corresponding members of Spanish scientific societies, and membership of international unions was very actively sought. By the 25th anniversary of the CSIC, in 1964, Spain was represented in every one of the fourteen scientific Unions belonging to the International Council of Scientific Unions – as its president, Oxford professor Harold Warris Thomson, pointed out as the celebrations opened in Madrid.[6] It must have come to no surprise, then, that in 1956 the CERN member states agreed to Spain and Austria's potential entry.[7] When CERN's first accelerator, a 600 MeV Synchrocyclotron, started providing beams of high-energy particles in 1957, the president of CERN's council, French diplomat François de Rose, who had served in Spain between 1953 and 1956, extended an invitation to

---

[5]   *Memoria de la Conferencia de Ginebra sobre usos pacíficos de la Energía Nuclear*, September 1955 (Archivo General de la Administración AGA 71/8470), as quoted in Romero and Sánchez Ron, 2001, p. 61.

[6]   'Speech by the President [of the International Council of Scientific Unions] to the Consejo Superior de Investigaciones Científicas on the occasion of their 25th anniversary on Oct 20 1964', Royal Society Archives, HWT/40/15.

[7]   In 1960, during the negotiations on Spain's entry, national delegates were asked to confirm 'the attitude that they had adopted in 1956 regarding an application by Spain for membership of CERN'; '23rd Meeting of the Committee of Council, 27 May 1960, Accession of Spain' (CERN/CC/358).

the Spanish government.[8] The economic situation of the country and the negotiation of the quota delayed entry until 1961 (Ceba and Velasco, 2012).[9]

Spain's contribution to CERN was subject to intense negotiations throughout the 1960s, showing the commitment of the Spanish government and Spanish physicists and officials to the organisation. Through the same period, national delegations at CERN were generally positive about Spain's membership and wanted it to stay. Indeed, contrary to the standard account, Spain left the organisation when a major reduction to their contribution had been secured, the biggest ever granted to a member state. This little known agreement, silenced at the time by the Spanish government in order to justify the exit, divided Spanish physicists.

Spain's contribution to CERN hovered above 50 million pesetas (some 2.7 million Swiss francs, around 4 per cent of the organisation's annual budget) between 1961 and 1963. Spain had also paid an initial special contribution of 2.7 million Swiss francs, down from 6.9 million (a 61.2 per cent rebate).[10] In 1963, the Spanish Delegation submitted a proposal to amend the method of calculating the contribution of member states, so that it would take into account the national income per capita, rather than gross national product and the stage of scientific and industrial development of each member. This prompted a revision of the criteria used as a basis for the calculation of contributions, which were compared with the criteria of other intergovernmental organisations. However, international statistics were 'very incomplete' and had not been collected on a comparable basis, while doubts persisted 'about the exact meaning of "scientific research"'.[11] The Finance Committee of CERN discussed the economic situation of Spain in great detail, building on data gathered from the International Monetary Fund and the United Nations and complementary information on the expenditure on education and research supplied by the Spanish government. The Committee noted that 'the first comprehensive study of research expenditure in Spain was made in 1963 in connection with the preparation of the Development Plan [1959]' and that there were 'no comparable data for previous years'. According to the figures provided by the

---

[8]   'El CERN y España', anon, n.d., p. 8 (GIFT Archive, Centre for History of Science, Universitat Autònoma de Barcelona); cf. Romero and Sánchez Ron, 2001, p. 88.

[9]   The Council of Ministers settled on entry in July 1960. *Boletín Oficial del Estado* (BOE), 175, 24 July 1961, p. 10998.

[10]   '33rd Meeting of the Finance Committee. Special Contribution from Spain', 13 June 1960 (CERN-FC-430).

[11]   J.M. Aniel-Quiroga to V.F. Weisskopf, 6 June 1963; '50th Meeting of the Finance Committee', 26 February 1963 (CERN-FC-602).

Spanish government (table 6.1), the expenditure on research and development was 1,513.8 million pesetas, which were very unevenly distributed between basic research (164 million, 10.8 per cent), applied research (719 million, 47.5 per cent) and development (630.8 million, 41.7 per cent). The distribution by scientific and economic sectors (table 6.2) showed the importance granted to Nuclear Energy, Biological and Agricultural Research, and Geology and Mining, which together got 712 out of 1,194 million pesetas (or 2/3 of the expenditure).[12] In December 1963, the CERN Council granted reductions in the Spanish contribution of 50 per cent in 1964, 35 per cent in 1965 and 20 per cent in 1966 (Romero and Sánchez Ron, 2001, p. 218).[13]

Table 6.1   Data concerning expenditure on basic research, applied research, development and services (in million pesetas)

	Basic Research	Applied Research	Research Total	Develop-ment	Services	Total
Centres and universities	162.0*	587.0	749.0	412.8	290.8	1,452.6
Industry and private sector	2.0	132.0	134.0	218.0	–	352.0
TOTAL	**164.0**	**719.0**	**883.0**	**630.8**	**290.8**	**1,804.0**

Note: * Including 50 million for contribution to CERN

Source: CERN Finance Committee, Fifty-fifth meeting, Geneva, 13–11–1963. (CERN/FC/639/Rev, annex II, p. 2). The data were provided by the Spanish government.

Table 6.2   Estimate of research expenditure in public establishments for some scientific and economic sectors

	PTA (millions)	% of total
Mathematics and Astronomy	14.0	1.17 %
Physics *	94.0	7.87 %
Chemistry **	54.6	4.57 %
Metallurgy **	45.0	3.77 %
Geology and Mining ***	162.4	13.60 %
Coal	13.0	1.09 %

---

[12] '55th Meeting of the Finance Committee, 13 November 1963', p. 4 (CERN-FC-639-Rev).

[13] '33rd Session of the Council, 14 and 15 December 1963. The Contribution of Spain' (DIR-ADM-FIN-06 (3) File 2).

	PTA (millions)	% of total
Biological Sciences and Agriculture	249.6	20.90 %
Building and Public Works	97.0	8.12 %
Fisheries	8.4	0.70 %
Medicine	17.4	1.46 %
Textiles	2.7	0.23 %
Aircraft Technology	92.8	7.72 %
Shipbuilding	6.0	0.50 %
Electrical Energy and Industry	2.0	0.17 %
Nuclear Energy	300.0	25.13 %

Notes

*   Including CERN contribution but not the nuclear physics research of the *Junta de Energía Nuclear* (JEN) nor the mechanics research of the *Instituto Nacional de Técnica Aeronáutica* (INTA).

**   Excluding the activities of the *Junta de Energía Nuclear* (JEN).

***   Including the activities of the Department of Geology and Mining of the *Junta de Energía Nuclear* (JEN).

Source: CERN Finance Committee, Fifty-fifth meeting, Geneva, 13–11–1963. (CERN/FC/639/Rev, annex II, p. 3). The column to the right has been added to show the percentage of each sector.

In 1966 Spain asked for a 35 per cent reduction of its contribution over the next triennium, arguing that it would otherwise contribute 7.1 per cent of the country's research and development annual budget (some 35 million dollars or 0.2 per cent of its GDP) to CERN alone. The Finance Committee was not willing to adopt a formula to calculate reductions on the basis of national income per capita, such as the UN formula $1/2 \cdot (1000 - x)/1000 \cdot 100$ – where x stands for average national income in US dollars and which results in 28.2 per cent for $x = 436$ USD, Spain's average income for 1962–64 – and instead recommended to follow an ad hoc procedure. The Council finally approved a reduction of 20 per cent for the triennium 1967–69, but even this proved untenable, and in a matter of months Spain was asking for a 28 per cent reduction, implicitly referring to the UN formula (table 6.3).[14]

---

[14] '79th Meeting of the Finance Committee', 18 November 1966 (CERN-FC-889); '80th Meeting of the Finance Committee', 13 December 1966 (CERN-FC-907); '33rd Session of the Council', 14 and 15 December 1966 (CERN-676); '35th Session of the Council', 21 and 22 September 1967 (CERN-735).

Table 6.3   Spain's contribution to CERN, 1961–71, in million pesetas

Year	Contribution	Reduction
1961	51.6	–
1962	58.1	–
1963	55.7	–
1964	25.0	50%
1965	40.3	35%
1966	81.7	20%
1967	133.9	20%
1968	156.2	20%
1969	130.7	50%
1970	136.5	50%
1971	161.1	50%

Note: The increase in 1968 reflects the devaluation of the currency (16%). The quantities for 1969–71 were estimated according to the reduction approved by the CERN Council in October 1968. *Source:* Lloret, 1968, p. 27.

While existing accounts suggest withdrawal was a clear-cut decision, dictated by the economic situation, the CERN archives actually show it was a protracted affair, much aired in the press. In August 1968 Spain expressed her intention to leave the organisation unless a 'substantial reduction' for the coming five years was granted. Early in October the CERN Council accorded a 50 per cent reduction in the Spanish contribution for the next three years, but this had no effect on the Spanish position. The press discussed the negotiations in explicitly autarkic terms. Thus an editorial comment titled 'No to autarky': 'How can we possibly cut the few ropes that tie us to Europe when our long term aim is to join the European Community? Autarky is a solution that leads to suicide and endemic underdevelopment and that distances us from the community of advanced nations'.[15] The Nuclear Energy Board issued a press release denying it had made public the official position of the government and rebutting 'all the comments and news on this matter that have of late been attributed to this Board by some newspapers'.[16]

The Council of Ministers agreed on 8 November 1968 to leave CERN and to devote part of the contribution to high energy physics (Gámez, 2012). The decision prompted student protests in the universities of Madrid and Barcelona,

---

[15]   *Diario de Barcelona*, quoted in *La Vanguardia Española*, 8 October 1968, pp. 10 and 13.

[16]   Press release, 22 October 1968, as included in *La Vanguardia Española*, 24 October 1968, p. 11.

while the press campaign in favour of permanence intensified: between July 1968 and February 1969 there were more than 150 press articles on CERN in Spain, 50 of them alone in November 1968.[17] *Cuadernos para el diálogo*, a dissenting journal tolerated by the Regime, published an anonymous anti-withdrawal article we now know to have been written by high-energy physicist Antoni Lloret (1968), based on a confidential report leaked by Otero Navascués.[18]

Whether due to this pressure or not, in December 1968 Spain asked the notice of withdrawal to be kept in abeyance and 'to remain "de jure" a member of the Organization, whilst accepting not to exercise her rights'. The letter, signed by Enrique Pérez-Hernández, the Spanish Permanent Delegate in International Organizations, could not be included in the agenda of the 40th Session of Council. Nonetheless, the Council took note 'with great satisfaction' of the declaration by the Spanish Delegate and postponed any decision until its next session.[19] In March 1969, Pedro Pascual, a rising Spanish theoretical physicist, was 'quite optimistic' that the situation could be reversed. But a couple of weeks before the Council session of June 1969, Spain communicated its final decision to withdraw with effect from 31 Dec 1968 'due solely to economic and financial reasons'.[20]

Meanwhile, other options had been considered, including remaining as an observer member. The French ambassador in Spain reported to the French Foreign Minister about the Franco Parliament session in which the issue was raised, following a battery of questions by deputy member, M. Trías Bertrán:

---

[17] Press Office, Nuclear Energy Board, 'Noticias sobre el CERN, no incluidas en otros boletines monográficos, aparecidas en la prensa española. 17 de julio de 1968 a 15 de febrero de 1969' (A. Lloret Papers, CEHIC).

[18] 'Informe de J.M. Otero Navascués', 11 July 1968 (A. Lloret Papers, CEHIC).

[19] '40th Session of the Council', 18 and 19 December 1968 (CERN-839 and CERN-843). '41st Session of the Council', 19 and 20 June 1969 (CERN-880): 'In its letter of 17 December 1968, the day prior to the start of the Fortieth Session of Council, the Spanish Government asked the Council to explore all the possibilities which might allow Spain to remain a member of the Organization, even if these might involve exceptional concessions. At the same time, Spain asked to remain "de jure" a member of the Organization for 1969, whilst accepting not to exercise her rights. In view of the fact that this letter had been received too late for Delegates to seek the views of their Governments on the matter, the Council decided to examine the question again at its Forty-first Session, scheduled to be held in March 1969'.

[20] P. Pascual to B. Gregory, 4 March 1969 (DIR-ADM-FIN-06 (3) File 3); '41st Session of the Council', 19 and 20 June 1969 (CERN-880).

We should like to know whether there are other reasons, besides the budgetary ones already announced, that justify Spain's decision to withdraw from CERN. At a time when the general consensus is that multinational aggregations are most convenient when it comes to this type of scientific research, will not the isolation from CERN put our scientific research at risk? Can we be sure that Spain will be able to sustain its own scientific research and development and that it will not spend more on royalties or patents than it will now save by withdrawing from CERN?[21]

The government's answer annoyed CERN officials. The official statement took for granted that Spain would have an observer status at CERN, even though no agreement had been reached on this matter. It also emphasised that the industrial development of nuclear energy in Spain and its short term economic benefits were to be preferred to a long term investment in international high-energy physics. The report by the French embassy in Spain reached CERN's Director of Administration, G.H. Hampton, who forwarded it to the General Director: 'the official statement ... totally ignores the CERN council offer of October. Do you wish me to do anything?'[22]

Spanish high-energy physicists were not convinced either, with the notable exception of C. Sánchez del Río, who in February 1969 claimed in the Spanish press that remaining in CERN would only strengthen the situation of the privilege of high-energy physics. A few months earlier the Professor of Thermodynamics at the University of Madrid, José Aguilar Peris, had already declared CERN to be 'a ruinous deal for Spain'.[23] Otero Navascués accused Sánchez del Río of silencing the truth and causing 'great damage'.[24]

Spain's subsequent attempts to secure observer status were met reservedly. In 1971 the Spanish Delegate, E. Pérez-Hernández, discussed with G.H. Hampton 'future relationships between Spain and CERN' and 'the state of the 300 GeV project'. In the case of application for admission:

---

[21]   *ABC*, 2 February 1969, p. 27.

[22]   'Respuestas de los ministerios de Educación y Ciencia y de Industria, al Procurador Señor Trías Bertran', *ABC*, 2 February 1969, p. 27; see also *La Vanguardia Española*, 1 February 1969, p. 4; Hampton to CERN's General Director, n.d. (DIR-ADM-FIN-06 (3) File 3).

[23]   C. Sánchez del Río, 'Puntualizaciones en torno al CERN', *Diario SP*, 14 February 1969, and *Diario SP*, 'El profesorado justifica la retirada de España del CERN', 6 February 1968, respectively.

[24]   Otero Navascués to Sánchez del Río, 17 February 1969 (A. Lloret Papers, CEHIC).

two points would be raised: their contribution to the capital costs incurred in the Organization between the time of their leaving in December 1968 and the time of their readmission and whether they would enjoy any reduction in their contribution. (My view is that though Council might waive part, if not all, of the capital costs, they would be extremely unlikely to accept Spain with a reduced contribution).[25]

In the early 1980s, when Spain sought readmission, these issues surfaced again. Spain was deemed to be 'very "fussy" concerning its contribution to the budget and always asks for an "extrawürst"'. Spain was also 'adamant where "juste retour" for its industries is concerned [...] and quite demanding in terms of positions within the organization'. As we shall see, these were also contentious issues back in the 1960s.[26]

Benefits: Staff and Contracts

Contribution meant little in itself. Beyond its huge symbolic value, the cost of belonging to Europe's leading scientific organisation had to be gauged against tangible benefits, including, above all, staff and contracts.

Used to participating in international organisations as a national representative, Otero thought that Spanish physicists and engineers had to be hired on the principal basis that Spain was a member of CERN.[27] His views ran against the organisation's policy, which placed professional skill and scientific achievements over nationality. Tensions came to the fore quite soon and they paradoxically involved CERN's first and higher-rank Spanish staff physicist, Rafael Armenteros (1922–2004), a Spanish exile who had graduated in Physics at Imperial College London in 1946. In December 1960, just a few days before Spain became an official member, Armenteros was offered a job at CERN. The problem with Armenteros was not just his Republican past, but

---

[25]   G.H. Hampton to CERN's Director-General, 26 January 1971.

[26]   Amb Pictet to CERN's Director-General, 8 March 1982 (CERN-ARCH-DG-HFS-0325).

[27]   Between 1950–66, Otero presided over the Spanish National Council for Physics (*Consejo Nacional de Física*, CNF). He enjoyed international recognition as national representative of Spain in organisations such as the International Union of Pure and Applied Physics (IUPAP), the International Committee on Weights and Measures (CIPM, which he presided between 1968–76), and the International Atomic Energy Agency (IAEA, whose General Conference he presided in 1971). See the hagiographical memoir Villena (1984).

also that Spain wanted to have a say in the recruitment of its own nationals. The Spanish Permanent Delegate in International Organizations, José Manuel Aniel-Quiroga, even suggested that staff members who they did not regard as Spaniards (that is, holders of a valid passport) should be listed as 'stateless', clearly referring to Armenteros.[28]

Shortly after joining the organisation, and following a suggestion by the Director-General, J.B. Adams, that two or three physicists visit CERN 'to study the work going on at our Laboratory and the way in which other Member States collaborate with us', Otero set up a delegation 'to discuss the possibilities for Spanish personnel to join the CERN'.[29] CERN had meanwhile advertised posts for engineers, physicists, mathematicians and operators in the Spanish press. Even though the Spanish Ministry of Work had approved the insets, Otero was not pleased that the JEN had not been told, and asked CERN's president Jean Willems not to take any decision until the Spanish commission had visited CERN.[30]

Less than three months into its membership of the organisation the first tensions surfaced, which the Spanish delegation tried to smooth over. One of the main Spanish dailies reported about the visit to CERN of five Spanish scientists 'with the aim of studying the possibility of integrating Spanish men of science within the organization'.[31] In his report about the visit, Samuel A. Dakin, CERN's Directorate Member for Administration, noted that:

> We had none of the anticipated trouble, in the sense that they did not take up the question of methods of recruitment, nor make any mention of Armenteros, except to inquire whether he was yet in post. But they had obviously come with an extremely clear and firm idea as to what Spain was going to get out of CERN, which was at this stage a suitable training for as large a number of her scientists as possible ... They reiterated these suggestions after they had made a round of Divisions, when apparently they found a good deal of interest in some of their possible candidates, I presume mainly on the assumption by the people they

---

[28] S.A. Dakin, *memorandum* of talk with J.M. Aniel-Quiroga, 10 January 1961 (DIR-ADM-PERS-02). Armenteros worked at the Track Chamber Division, directed by Charles Peyrou. Both men had met at Louis Leprince-Ringuet's laboratory at the École Polytechnique (Baillon, 2005; Mersits, 1990, pp. 178–9).

[29] Otero to J.B. Adams, 18 January 1961 (DIR-ADM-PERS-02).

[30] Otero to J.B. Adams, 11 March 1961 (DIR-ADM-PERS-02); *ABC*, 23 February 1961, p. 60; *La Vanguardia Española*, 23 February 1961, p. 13.

[31] *La Vanguardia Española*, 28 March 1961, p. 13. The members of the Spanish delegation were C. Sánchez del Río, J. Catalá, María A. Vigón, A. Tanarro and F. Verdaguer.

spoke to that the Spaniards would be fellows or would come outside existing staff ceilings ... We promised them nothing, emphasized the competition for fellow and staff posts, but said that we would do what we could to meet the situation.[32]

Later that year Otero expressed his views that Spain's 'economic sacrifice' should be rewarded with a group of researchers belonging to CERN's staff and put forward Joaquín Catalá (1911–2009), a senior Spanish physicist, for a permanent position. Director-General John B. Adams told Willems that the candidate's experience was 'not sufficient in high energy physics to justify a grade which would be acceptable to him' and instead offered a 'visiting scientist' position although 'he hardly qualifies scientifically for [it]'. His judgement was based on an appraisal by G. Bernardini: '[the candidate's] papers show a standard physicist. Some of them approach technical problems rather lightly, others give the impression of disproportion between the amount of work and the data and conclusions derived from them'. Adams thought a point could be stretched in this case as long as it was not taken as a precedent and Willems's final answer was very carefully worded so as to make the post appealing to both Otero and his candidate. I take this exchange to be extraordinarily revealing of the rather different views on recruitment and scientific excellence held by Otero and CERN's scientists and officials.[33]

The issue was taken up over and over again in the following years, both parts being concerned about the 'disproportionately small' number of Spanish staff at CERN. Over the first 18 months of membership, about 40 Spanish candidates were called to selection boards, but the JEN's focus on applied nuclear sciences appears to have backfired, because 'although many of the Spanish candidates have experience in nuclear physics or engineering, this is very commonly in reactor work which has little application to the sort of work we do here'.[34] By October 1962, however, there were eight Spanish staff members, including four scientists or engineers.[35] By mid-1963 the situation of Spain was not unlike that

---

[32]   S.A. Dakin, 'Note of the Talk with the Spaniards', 30 March 1961 (DIR-ADM-PERS-02).

[33]   Otero to Willems, 15 June 1961; Adams to Willems, 29 June 1961; Bernardini to Adams, 29 June 1961 (DIR-ADM-PERS-02).

[34]   Dakin to Otero, 3 October 1962 (DIR-ADM-PERS-02); see also Otero to Weisskopf, 5–09–1962, and G. Ullmann to S.A. Dakin, 17 September 1962 (DIR-ADM-PERS-02).

[35]   Ullmann to Dakin, 17 September 1962 (DIR-ADM-PERS-02) includes a staff list. The eight people with the higher grade were men, the six people of lower grade were women: R. Armenteros (*Senior Physicist, Grade 13*); D. García Fresca (*Senior Engineer, Grade 11*); A. García González (*Mechanical Engineer, Grade 9*); J. Goñi Unzué (*Physicist, Grade 9*); A. de

of Sweden, a founding member of the organisation, which contributed 4.18 per cent of the budget (as compared with Spain's 3.36 per cent) and had 13 members of staff, including so-called supernumeraries (as compared with Spain's 16).[36] Spain tried all along, with limited success, to retain some control of the recruitment process, having potential candidates send their CVs to the Spanish Ministry of Foreign Affairs or the Spanish Institute for Emigration, rather than to the organisation itself, as was usually the case.[37]

The clash of research cultures was also apparent when Spanish physicists trained abroad sought to return. From 1967 an Experimental Group of High-Energy Physics operated at the Nuclear Energy Board quarters in Madrid, under the direction of Antoni Lloret, a Spanish physicist who had done his PhD in France. Lloret recruited a small group of French and Spanish physicists (Bruno Escoubès, Miquel Tomàs and Salomé de Unamuno) that scanned bubble chamber photographs with hardware and software borrowed from CERN or acquired in France. The creation of the group was personally supported by CERN's Director-General B. Gregory and intended to buttress Spain's membership. However, the group faced internal opposition and disbanded in 1969–70, to be reconstituted on a different basis shortly thereafter. Local physicists worried about 'brain gain', while the Rector of the Universidad Autónoma de Madrid and future minister of Education and Science, Julio Rodríguez Martínez, praised the 'non-drained brains' (*cerebros no recuperados*) that had reconstructed the Spanish university since the end of the Civil War and were now opening it 'to those who left'.[38] In 1977 Lloret would still insist, at a meeting to discuss science policy in high-energy physics, that 'it should be kept in mind that today it is inconceivable to do autarkic research, not just in relation to European countries but also in relation to the different countries that make the Spanish state' (Morales, 1977, p. 96).

Together with staff underrepresentation, the lack of contracts with Spanish companies also played a role in the decision to withdraw from CERN. How were these seen from the inside, as far as CERN sources allow us to tell? 'Failure

---

Tena (*Tech. Draughtsman I, Grade 6*); J. Acosta Sánchez (*Electrician II, Grade 5*); L. López (*Mechanic II, Grade 5*); R. Paredes (*Operator II, Grade 5*); M. Escribano (*Operator II, Electr. Comp., Grade 4*); M. Morales Kamenski (*Clerk II, Grade 4*); B. Goyenechea (*Scanner, Grade 3*); C. Goyenechea (*Operator IEP, Grade 3*); T. de Julián (*Lab. Handywoman, Grade 3*); C. Prado (*Operator IEP, Grade 3*). A list dated 21 February 1963 includes another woman physicist, M. Guinea-Moorhead (*Physicienne, Grade 7*).

[36] 'Ingenieurs et physiciens espagnols au CERN', 2 August 1963; 'List of staff 1961–64' (DIR-ADM-PERS-02).

[37] J.A. Giménez-Arnau (Spanish ambassador in Geneva) to G.H. Hampton, 10 May 1965; Hampton to Giménez-Arnau, 24 May 1965 (DIR-ADM-PE-02).

[38] J. Rodríguez, 'Cerebros no recuperados', *ABC*, 8 May 1973.

in communications' between CERN and Spanish industry was suggested by S. Dakin and rejected by Otero.[39] The results of Spanish firms were certainly disappointing: by 1965 they had been awarded contracts for 25,545 Swiss francs, a negligible percentage of the 452 million Swiss francs total of contracts awarded between 1952 and 1965. However, firms in other countries did not fare much better. While CERN's policy was based on 'competitive tendering' (whereby firms were invited to submit offers), rather than 'just return', the gulf between principle and practice meant that not all contracts were competitive: submitting an offer was a costly process that was not compensated if the firm failed to win the contract. As Krige has shown, the process was neither completely rational nor objective, but it was maintained mostly for political reasons, as it represented:

> an agreement between the member states that no consistent effort be made to distribute contracts between them in relation to their contributions to the budget. The *practical implications* of the 'competitive' policy are thus best understood *negatively*, as meaning that there is no policy of just return in the award of contracts for plan, equipment, and supplies (Krige, 1990, pp. 671 and 668; Krige, 1987).

Conclusion

Sources at CERN provide a new perspective on Spain's first membership of the organisation. Political will was not lacking, but was rather rooted in autarkic practices and attitudes. An illuminating event was Spain's candidature for the site of the new 300 GeV Super Proton-Synchrotron that CERN began planning in 1964 and finally built in Geneva. The Spanish delegation chose a site carefully and settled for a place that had 'all desirable geological conditions' and was conveniently close to 'Madrid and Barajas international airport'. As it happens, El Escorial was also the site of Philip II of Spain's palace and monastery, the ultimate symbol of the political and religious power of the Spanish Empire. In the inaugural speech of the CERN School of Physics that took place at El Escorial between 26 May and 8 June 1968, before Queen Frederica of Greece, the Reverend Father Prior and a select group of European physicists and Spanish authorities, Otero referred to El Escorial 'as a sort of Spanish shrine' and 'one of the milestones of our power', built by 'the very first King of Spain ... and the

---

[39]   Dakin to Otero, 3 October 1962 (DIR-ADM-PERS-02); Otero to Dakin, 23 October 1962 (DIR-ADM-FIN-06 (3) File 1).

first modern king who is fond of science'. He went on to depict in bullfighting terms the transformation of the estates where the Big Machine was to be built: 'We assume that where bulls run now, protons could run as well, though at a greater speed'.[40] Otero's anachronistic speech was at odds with his efforts to save Spain's membership, which failed to change the mind of those who had granted him international status. National representation had granted Spain a place in many an international organisation and provided a Navy engineer such as Otero with scientific credentials. Quotas had effectively worked in many instances and would work with the European Space Research Organization (ESRO), which Spain nearly left at about the same time; but they did not work with CERN, which was run on different principles (Sánchez Ron, 1994). The failure of Spain's first stage at CERN may have been a failure of autarkic physics.

In 1983 Spain became again a member of CERN, in time to join in the celebrations of the organisation's 30th anniversary. In his commemorative speech, King Juan Carlos I of Spain quoted George Sarton: 'Scientific work is the result of an international collaboration, the organization of which is perfected every day. Thousands of scientists devote their whole lives to this collective work like bees in a hive but their hive is the world'. Reporting on the ceremony, a physics professor at the Universitat Autònoma de Barcelona pointed out that the future of science in Spain did not just hinge on the number of scientists and the budget for research, but rather on the creation of 'a proper atmosphere for the conduct of science'. Aware of it or not, he was probably providing the best explanation of Spain's failed attempt, in the 1960s, to consolidate as a member of Europe's major scientific organisation.[41]

List of References

Aguilar Benítez, M., 2004. 'CERN 50 aniversario'. *Revista Española de Física*, 18(4), pp. 3–15.

Aguilar Benítez, M., 2007. 'El CERN'. *Revista de la Real Academia de Ciencias Exactas, Físicas y Naturales*, 101(1), pp. 183–201.

---

[40] 'Speech by Professor Otero Navascués at the Inaugural Meeting of the CERN School of Physics at El Escorial', p. 5 (CERN archives, CERN-ARCH-SCSA-094).

[41] Ramon Pascual, 'Treinta años de una empresa común', *La Vanguardia Española*, 2 October 1984, p. 5. The king's speech is quoted in Aguilar (2007), pp. 183–201. Sarton's quote is given in Spanish without a reference; I have quoted from Sarton (1948), p. 56, which comes very close to Aguilar's rendition.

Aguilar Benítez, M. and Ynduráin, F.J., 2003. 'El CERN y la Física de altas energías en España'. *Revista Española de Física*, 17(3), pp. 17–25.

Anduaga, A., 2009. 'Autarchy, Ideology, and Technology Transfer in the Spanish Oil Industry, 1939–1960'. *Comparative Technology Transfer and Society*, 7(2), pp. 172–200.

Baillon, P., 2005. 'Rafael Armenteros, 1921–2005'. *CERN Courier*, 4 July 2005.

Camprubí, L., 2014. *Engineers and the Making of the Francoist Regime.* Cambridge, MA: The MIT Press.

Claret, J., 2006. *El atroz desmoche. La destrucción de la universidad española por el franquismo, 1936–1945.* Barcelona: Crítica.

Ceba, A. and Velasco González, J., 2012. 'The Entry of Spain into CERN during Francoism in 1961: Valencia's Role'. In: A. Roca-Rosell (ed.), 2012. *The Circulation of Science and Technology: Proceedings of the 4th International Conference of the ESHS.* Barcelona: SCHCT-IEC, pp. 748–53.

Gámez, C., 2012. 'La física teórica de altas energías en España durante la dictadura del general Franco'. In: N. Herran and X. Roqué (eds), 2012. *La física en la dictadura. Físicos, cultura y poder en España, 1939–1975.* Bellaterra: Servei de Publicacions UAB, pp. 141–57.

Gómez, A. and A.F. Canales (eds), 2009. *Ciencia y fascismos. La ciencia española de posguerra.* Barcelona: Laertes.

Guirao, F., 1998. *Spain and the Reconstruction of Western Europe, 1945–57: Challenge and Response.* London and New York: Macmillan/St. Martin's Press.

Herran, N. and X. Roqué, 2013. 'An Autarkic Science: Physics, Culture, and Power in Franco's Spain'. *Historical Studies in the Natural Sciences*, 43(2), pp. 202–35.

Kaiser, D., 2005. *Drawing Theories Apart. The Dispersion of Feynman Diagrams in Postwar Physics.* Chicago: University of Chicago Press.

Krige, J., 1987. 'Which firms got CERN contracts?'. *Studies in CERN History* (CHS-24).

Krige, J., 1990. 'The contract policy with industry'. In: A. Hermann, J. Krige, U. Mersits and D. Pestre (eds), 1990. *History of CERN. Vol. 2, Building and running the laboratory.* Amsterdam; Oxford: North-Holland, pp. 637–77.

Krige, J. (ed.), 1996. *History of CERN. Vol. 3.* Amsterdam; Oxford: Elsevier.

Lloret, A., 1968. 'La retirada de España del CERN'. *Cuadernos para el diálogo*, 62, pp. 25–8.

Low, M., 2005. *Science and the Building of a New Japan.* Basingstoke: Palgrave Macmillan.

Malet, A., 2008. 'Las primeras décadas del CSIC: investigación y ciencia para el franquismo'. In: A. Romero de Pablos and M.J. Santesmases (eds), 2008. *Cien años de política científica en España*. Bilbao: Fundación BBVA, pp. 211–56.

Malet, A., 2009. 'José María Albareda (1902–1966) and the formation of the Spanish Consejo Superior de Investigaciones Científicas'. *Annals of Science*, 66(3), pp. 307–32.

Mersits, U., 1990. 'The construction of the 28 GeV Proton Synchrotron and the first six years of its scientific exploitation'. In: A. Hermann, J. Krige, U. Mersits and D. Pestre (eds), 1990. *History of CERN. Vol. 2, Building and running the laboratory*. Amsterdam; Oxford: North-Holland, pp. 139–267.

Morales, Á. (ed.), 1977. *Actas de la Mesa Redonda sobre Política Científica Española en la Física de Altas Energías*. Zaragoza: Departamento de Física Atómica y Nuclear, Universidad de Zaragoza.

Pascual de Sans, P., 1998. 'Los inicios de la Física de Altas Energías en España'. *Arbor*, 159(626), pp. 231–9.

Pestre, D., 1984. *Physique et physiciens en France, 1918–1940*. Paris: Éditions des archives contemporaines.

Pestre, D., 1990a. 'The CERN system, its deliberative and executive arms and some global statistics on how it functioned'. In: A. Hermann, J. Krige, U. Mersits and D. Pestre (eds), 1990. *History of CERN. Vol. 2, Building and running the laboratory*. Amsterdam; Oxford: North-Holland, pp. 341–415.

Presas, A., 2007. 'Las ciencias físicas durante el primer franquismo'. In M.Á. Puig-Samper (ed.), 2007. *Tiempos de investigación. JAE-CSIC, cien años de ciencia en España*. Madrid: CSIC, pp. 299–303.

Romero de Pablos, A. and Sánchez Ron, J.M., 2001. *Energía nuclear en España: de la JEN al CIEMAT*. Madrid: CIEMAT.

Romero de Pablos, A. and Santesmases, M.J. (eds), 2008. *Cien años de política científica en España*. Bilbao: Fundación BBVA.

Roqué, X., 2012. 'España en el CERN, o el fracaso de la física autárquica'. In: N. Herran and X. Roqué (eds), 2012. *La física en la dictadura. Físicos, cultura y poder en España, 1939–1975*. Bellaterra: Servei de Publicacions UAB, pp. 239–58.

Sánchez Ron, J.M., 1994. 'Poder científico versus poder político: reflexiones a propósito del CERN y de ESRO/ESA'. *Arbor*, 147(577), pp. 27–49.

Santesmases, M.J., 1999. *Antibióticos en la autarquía: Banca privada, industria farmacéutica, investigación científica y cultura liberal en España, 1940–1960*. Madrid: Fundación Empresa Pública, 1999.

Saraiva, T. and Wise, M.N., 2010. 'Autarky/Autarchy: Genetics, Food Production, and the Building of Fascism'. *Historical Studies in the Natural Sciences* 40(4), pp. 419–28.

Sarton, G., 1948. *The Life of Science. Essays in the History of Civilization*. New York: Henry Schuman.

Traweek, S., 1992. *Beamtimes and Lifetimes. The World of High-Energy Physicists*. Cambridge, MA: Harvard University Press.

Villena, L., 1984. José María Otero Navascués (1907–1983). *Óptica Pura y Aplicada*, 17, pp. 1–12.

Chapter 7

The National Council for Research in the Context of Fascist Autarky

Roberto Maiocchi

Introduction

This chapter examines the history of the National Council for Research (*Consiglio Nazionale delle Ricerche* – CNR), the most important Italian scientific institute of the Fascist years, taking the account up until the Second World War. The focus is on the role of the institute in carrying out the autarky project, which involved the whole Italian society from 1935 onwards.[1] It is argued that the National Council for Research (CNR) would eventually prove to be unable to reach the goals set for it by the political powers.

Founding the National Council for Research

The official proclamation of autarky – the fundamental fascist political project – appeared in a speech given by Mussolini at the end of the great manoeuvres in Bolzano on 31 August 1935. On this occasion, the Duce announced to the World that Italy 'would manage alone' (Maiocchi, 2003, p. 125). In fact, the question of economic autonomy for the nation had been discussed in scientific–technical circles for twenty years, that is, ever since the First World War revealed the weaknesses which endangered the foundations of the Italian economy.

The Great War was a turning point for the history of science in Italy. During those dramatic years, the scientific community and part of the political and economic world harshly criticised the existing way science was organised and proposed its thorough reformation. At the moment of its constitution, the united Italian State had inherited a huge number of universities and institutes

---

[1]   The present essay is based on some of my previous works, which will be cited throughout this chapter.

of higher education from the multitude of former small political unities, an abundance that was totally out of proportion to the economic possibilities of the State budget. These scientific institutes were each very small, and in most cases their human resources were limited to the professor who held the chair, to his assistant and to a janitor. As a compensation for the scarcity of means and personnel, the institutes' directors had absolute freedom of research and so did not answer to anybody for their work. This extremely time-wasting situation had given birth to scientific activity characterised by the absence of laboratories able to compete with the major foreign laboratories, and by the prevalence of pure science over applied science. The various governments that had followed one another until the war had tried many times to reduce the number of universities without ever succeeding, because of strong local opposition.

The war mercilessly exposed serious deficiencies in the production sector and a shortage of basic raw material – problems which other countries involved in the war (particularly Germany) had approached with the decisive help of applied science (Maiocchi, 2000, pp. 209–44). During the conflict an ideology of 'technical-scientific nationalism' was born, which, by means of conferences, publications, organisational initiatives and political pressure, drove attempts to concentrate resources (including the founding of large national research institutes) and to increase the involvement of Italian scientists in applied research (which created the basis for a better use of national resources) (Maiocchi, 1993). Those two points of the programme, namely the fight against the waste of resources and developing a science useful for the nation, provided the cultural background for numerous public and private initiatives. The greatest of these initiatives was the creation of a Bureau for Inventions and Research Initiatives by the Ministry of War, started in the first place by Vito Volterra, an internationally recognised mathematician (Maiocchi, 2000b and 1998)· The Bureau carried out a coordinating role, using the laboratories of universities or State organisations such as the National Railways, and using the competence of university professors. The Bureau grew progressively, reaching (formally, at least) a staff of almost sixty people.

By the end of the war, thanks to Volterra, a project aimed at creating a public institution detached from university circles was developed. Such a step would permit the opening of a large State laboratory (to be eventually divided into three laboratories, separately for physics, chemistry and biology) and push Italian science towards studying questions regarding the economic development and security of the nation. This institution was the National Council for Research. Although the project faced many difficulties originating in political circles, and many times it seemed that it would collapse, the National Council for Research was eventually founded on 18 November 1923 with Volterra as president (Simili and Paoloni, 2001, p. 15). The decisive political impulse that convinced Mussolini to approve the decree certainly did not come from Volterra, but from two political men, both fascists from the start and also well connected with the institutions: Amedeo Giannini, an official of high rank in the Ministry of Foreign Affairs, and Giovanni Magrini, who had worked inside different national and international institutions and who was in the centre of a thick web of personal relationships. Giannini became the vice-president and Magrini became the secretary of the CNR. The reasons that pushed Mussolini to approve the Council are not clear at all. His attitude during the following years arouses the suspicion that in 1923 Mussolini did not have a clear idea about what Volterra's project meant.

At that moment, the cultural atmosphere was changing. The economic post-war crisis was already over and what followed was a period of economic growth which, despite some slackening, continued until the great crisis of 1929. This economic growth resulted from an intensive international exchange which did not really fit into the aim of decreased imports. This idea of decreased imports had been prominent in the preceding years. Yet, within ten years, the scientific–technical debate over the possible economic autonomy of Italy, which had been excessively intensive at the beginning of the 1920s and concentrated particularly on nitrogen fertilisers and combustibles, occupied less and less space (Maiocchi, 2003, chapter 1).

The National Council for Research, based upon a quasi-autarky programme, remained practically frustrated. During the four-year presidency of the anti-Fascist Volterra, government subsidies were enough only to maintain its existence (175,000 liras per year which today would equal about 110,000 euros) (Maiocchi, 2000c and 2004, chapter 1). Besides keeping itself alive, the Council could not do much and just organised the publication of a Scientific–Technical Bibliography and maintained relationships with the International Council of Research, of which it was a member. In 1927, after Volterra's tenure of office had expired, the post was offered to Guglielmo Marconi, who had invented the radio. Marconi was a businessman who led a sort of multi-national social existence; he was a complete stranger in the Italian academic circles, chosen by Mussolini only because of the splendour he could add to the institution (Simili and Paoloni, 2001, chapter 2). During the handover of the post to Marconi, the Council underwent a restructuring – a work which lasted two years, so that its actual and real activity could only start in 1929. During this period, Marconi resided in Great Britain to run his business.

The strategic project that justified the Council's existence remained the same: the plan was to constitute a great national laboratory (or some big national laboratories), independent from the university world and destined to undertake research in applied science with the aim of developing Italy's economic potential. Still, the resources at the disposal of the CNR remained very small (679,000 liras a year, about 400,000 euros) which excluded the possibility of putting into practice even a part of the project to create research institutes of national character. As an answer to Marconi's many requests, Mussolini promised money in the distant future. Moreover, until 1937 the institution remained without its own head office: it was instead forced to be hosted at various public offices or rented private apartments. As a consequence, at times it had to face the necessity of dismissing some of its employees (Maiocchi, 2003, chapter 1).

Undoubtedly, this lack of generosity regarding the funding of the CNR resulted from Mussolini's attitude. At least until 1930, Mussolini had serious doubts about the utility of the CNR because he saw it, above all, as a propaganda instrument dedicated to organising conferences and exhibitions, issuing publications and popularising a perfect image of Italian science abroad. As such, to Mussolini, the Council seemed a useless copy of the Italian Academy – a representative body created specially in 1926 to glamourise the culture of Fascist Italy. So, Mussolini thought seriously about opportunities to liquidate the unit, and did not feel any need to charge the CNR with strategic research project directives (Maiocchi, 2003, pp. 26ff.).

Even the intention of shaping the CNR as an agency that was separated from the university world, which Magrini, in a memorandum from 1929, described as 'the environment which is traditionally the most hostile to the regime', had to face hard reality. There were no scientists, no good technicians and no laboratories outside the universities. Thus, the CNR management was made to appeal to the university world in order to be able to carry out research activities. Instead of dividing the scientific personnel into CNR and university members, it became a separation between the university professors that had managed to get into the CNR and those who were left outside. The entrance of the professors led to the creation of research projects that were inspired more by the interests of individuals than by general intentions. Some of these projects were also very successful, as in the case of Enrico Fermi's research on artificial radioactivity, financed by the CNR.

Awaiting a more convenient time, with absolutely no hint from the government, the CNR turned in an autarkic direction on its own initiative. In the narrow management group, particularly noticeable because of their influence, were the vice-president, Amedeo Giannini, a professional diplomat involved in science, and Nicola Parravano, Professor of Chemistry at the University of Rome, who was very involved with industry. In their vision, Italy appeared as a country whose economy should be based predominantly on agriculture and which was able to follow a path of economic development different from the model displayed by states where capitalism was already advanced (United Kingdom and the United States). In other words, which should develop through focusing on industrial production linked to agriculture. Only by being aware of this ruralist perspective can one understand the first research projects initiated at the CNR, almost all of which were focused on rational – direct or indirect – use of Italian agricultural resources. Great attention was granted to the use of wood as fuel, with the utilisation of gasogene material. Particular attention was paid to the processing of citrus fruit where, in accordance with the Institute for Export, the CNR managed to obtain a patent on a mechanical procedure for extracting lemon essence from lemon paste, the by-product of citric acid production. Among the other issues studied were glycerine production through fermentation of agricultural by-products, tomato conserves, mineral waters, producing ethanol from agricultural products and the use of castor oil as a 'national' lubricant (Maiocchi, 2001).

Economic Decline and the Mission of the CNR

In 1931, however, the political and economic climate in Italy started to change. It was only during this year that the seriousness of the international economic crisis was fully evaluated. Italy made efforts to maintain a liberal foreign trade policy even after such powerful countries as England had adopted protective measures. The Fascist government had to recognise that the economic problems could not be solved by turning towards the international market, and that the situation required some regulation of foreign trade and an increase in domestic production. This decision, which anyway was to be carried out *in toto* only within the following two years, was accompanied by the decision to finally put into practice the reorganisation of the Italian production sector by founding it upon, as had been discussed for some years, a corporatist model. At the same time, during the last months of 1933, Mussolini took the decision of beginning a policy of colonial conquest, preparing for the conquest of Ethiopia. Therefore, in 1933 a clear political line was drawn – a line that aimed at mobilisation of all the national resources, seeking for the greatest possible independence from abroad in view of preparing for conquest and war. It was the autarky project, even if the term itself was not yet in use.

In this new context, the ideology that I have defined as 'scientific–technical nationalism', which was at the roots of the initial project of the CNR, regained its power. The Council tried to adapt to the requirements imposed by the historical moment by launching certain initiatives meant to contribute to the economic independence of Italy. Particularly remarkable was the activity of the Committee regarding raw material used in Italian production (Maiocchi, 2001, p. 35). The president of the Committee was Gian Alberto Blanc, a chemist who was deeply involved in various industrial initiatives. In his speech delivered at the opening of a plenary reunion of the CNR on 7 March 1933, Marconi confirmed that the raw material issue was the central point of the Council's programme. The following year, talking at the plenary reunion held on 8 March 1934, Marconi came up with what could be called an innovation when compared to his previous public appearances; innovative because he combined the familiar subject of the value and exploitation of national resources with the theme of the imperial mobilisation of science (Maiocchi, 2001, p. 39).

Mussolini seems to have decided that Italian science would be involved in military preparations: he ordered a considerable increase of funds for the CNR. While between 1930 and 1934 its average funding was about 1,500,000 liras (ca. 1,200,000 euros), in 1935 almost 6 million liras (5,300,000 euros) was assigned to the Council (Maiocchi, 2000a). So the money assigned was four times as much as in the previous years. On 18 May 1934, the Duce approved the order which constituted the Co-ordination Committee between the CNR and the army. The Committee's first session, presided by Marconi, took place on 9 July 1934. Records of this Committee have been lost, but most probably its activity did not continue beyond 1934. We do know, however, that its inauguration session was conducted in a truly solemn manner (Maiocchi, 2003, p. 39). In his speech, Marconi described a country already at war: 'When a people gets to fight with another people, all its energies aim at victory, because it is a matter of life or death, and science has the duty to give the courage of men the support of the best weapons and of the most intelligent resources of technique'.

Regardless of numerous official declarations, none of the Government's representatives seemed to consider the CNR as a useful consulting body. The ministers and the army preferred to address their own technical offices and evidently considered the CNR a rival of which to be jealous, rather than an instrument of technical and scientific information. The CNR, although it did not have necessary strategic information, had to decide alone which problems were most urgent to work on. On 6 March 1935 Mussolini sent a letter to Marconi in which he indicated problems which, in view of the war, should further be considered fundamental in the final stage of realising economic autarky.[2] At that time, the preparations for the war in Ethiopia, which was to start in October of the following year, were already in full progress.

The Duce posed four fundamental questions and asked the CNR to swiftly resolve them:

> It is absolutely necessary that the CNR should concentrate its efforts on the following problems in order to find both a national and an industrial solution to them (that is, not just on a simple laboratory level). A) the national fuel problem (alcohol, rocks and schist, gas generators etc.); B) the national fabric problem [in time this prepared the way for mixing hemp with cotton]; C) the national cellulose problem; D) the national problem of the utilization of solid combustibles (coal, brown coal etc.). On some of the problems listed there are studies, experience and industrial applications (in initial stages). It is time to give the Government the grounds to act on a wider scale.

The problems brought up by the Duce, as well as other issues, had been discussed for a couple of months by the press, but the CNR did not take them into consideration except for the 'cottonization' of hemp (mixing cotton with fibres made of hemp): for this purpose they rented a laboratory in a technical institute in Naples and left it at the disposal of the putative 'inventors'. It was all about enormous problems to which there seemed to be no quick solutions and which could only be reasonably approached if one had much time and vast resources. Mussolini did not concede either to the CNR, however his directive could not be ignored. The CNR reacted rapidly and within less than two months the reports expected by the Duce were ready (Maiocchi, 2003, p. 50). Talking about the cottonization of the hemp, it was explained that the laboratory in Naples was waiting for someone to turn up with some ideas to test. As for the cellulose, there was a summary of five pages about the debate in the newspapers. About the national fuel and combustibles problem, some old reports were attached, with the warning that they were 'provisory' reports and that there was need for money and time to get to definitive solutions. Of course, as might have been supposed, the reports were absolutely useless and sank into oblivion. Mussolini never personally asked the CNR again for anything.

Also in spring 1935, another important sign was given of a modest growth of interest by the Government in the activity of the CNR, namely the creation of the Inter-Ministerial Commission for Insufficient Raw Material and for

---

[2]   On *Archivio Centrale dello Stato*, Roma, ACNR (Archivio CNR), b. 981, II.

Substitutes (CISS). This was the unit tasked by the Supreme Commission for Defence, the highest governmental body with military prerogatives, whose head was Mussolini (Maiocchi, 2003, chapter 4). The task of the Commission was of great strategic relevance. The Commission was to issue a report in January of the following year. The report, which was to be presented to the Supreme Commission for Defence, was supposed to indicate the needs, the effective resources, the deficits and the way to obviate the problem of the possibly broad variety of raw material Italy would need in a hypothetical first year of war. In other words, the report was to contribute to the evaluation of whether the Nation was able or not to cope with a year of war. The Commission was meant to be permanent and to prepare this kind of evaluation every year; as such it represented the most important form of involvement of the CNR in the war preparations.

The Commission comprised technicians representing various ministries, the Armed Forces and the CNR. The latter was also supposed to provide the head office and contribute to organisational needs. Since the CNR did not possess any unit for publishing statistical data (the data included in Blanc's report regarding raw material would later turn out to be unreliable), basic statistics for the report on the necessity and availability of raw material were requested from the Committee for Civil Mobilization, a military structure created during the First World War to manage the production of economic goods in case of war. The head of the Committee was General Alfredo Dallolio, a tough and elderly officer, who would always make it evident that he considered the CISS as a useless and annoying, if not harmful, rival. During the first meeting of the CISS, on 21 May 1935, Dallolio had his representative, Colonel Aurelio Cossi, say that his Committee did not intend to let the CISS access the data it possessed, because it was a state secret. This was a not too elegant way to tell the members of the CISS, who themselves were obliged to abide by the secrecy, that they were considered as chatterers, if not as real spies. Only after considerable pressure did Dallolio resign himself to communicate the requested data, although he continued to maintain an irritated and un-cooperative attitude. The CISS had to get used to work on the basis of these features, sometimes correcting and sometimes making proposals. Still, regardless of these difficulties, the CISS pursued its activity in the following years and would issue its annual reports on time. The Commission was gradually broadened: outstanding scientists and technicians from private industry were employed, the work was divided and articulated efficiently, and so the scale, precision and concreteness of the final reports increased noticeably.

It seems that Mussolini paid much attention to the CISS reports, but – unfortunately – also that the CISS paid much attention to Mussolini's opinion.

In their final discussion about the preliminary works, one can sense a growing worry not to provide an excessively negative picture of the situation in Italy, smoothing the available data in order not to cheat the expectations of the Duce. In the execution of this preventive censorship, Amedeo Giannini, the vice-president of the CNR, was particularly active. The evaluation of pit coal included in the report from January 1940, and which I describe next, seems to be the most ostentatious example of 'mending' data to support Mussolini's strategic choices instead of confronting them with reality.[3]

January 1940 was a particularly dramatic period: several months before, Europe had fallen prey to the advancing Wehrmacht and Italy had to decide whether it should enter the war as Hitler's ally or not. The CISS report was to serve as a reference point for an epochal decision in Italian history. The report contained disconcerting data on pit coal. It represented the most important import item, reaching about 13 per cent, in value, of Italian imports. The amount of imported coal gradually increased and in the years 1938–39 it exceeded 12 million tonnes. The 1940 CISS report contained both a clear and surprising suggestion: if, in case of peace, the need for combustibles to import was expected to be 12,750,000 tonnes, in case of war the estimation was reduced to 8,900,000 tonnes, that is, the amount guaranteed in secret agreements with Germany. Thus, it meant that, in the case that Italy entered the war, the country's needs for coal consumption would be reduced. The miracle of reducing consumption by almost 4 million tonnes of combustibles would have been put into practice by means of a drastic decrease of industrial production: for the Italian industries during peacetime there was a reserve of 9.5 million tonnes, but in case of war industry would have to do with less than 6.5 million tonnes, about 4 million of which was destined to the war industries. It seems more than evident that such a solution, based upon the almost complete paralysis of industries which were not meant to be exploited in war, could only be seriously considered in case the war would not last long; indeed, the country could survive and fight only for a few months with its industrial structures barely working, or even out of work, to avoid consuming coal. As far as combustibles are concerned, the decision to enter the war seemed to be a bet, a great risk that could only be taken into consideration if one had forgotten all that was written and said about the principal conclusions that should have been drawn during the two preceding decades from the experience of the Great War: modern war was no longer a war of armies but a war of nations which required the complete involvement of all

---

[3]   CISS, *Quinta relazione a lla Commissione suprema di difesa 1940. Combustibili fossili,* on *Archivio Centrale dello Stato,* Roma, Archivio CNR, b. 264.

the productive forces, the maximisation of industrial activity, but certainly not its slackening. To take this risk by trusting in a swift solution to the conflict was a dramatic step. Mussolini, though, chose the risk and the CISS report provided data which were mostly welcomed by the Duce. Immediately after Italy had entered the war, the CISS was dissolved because of the fact that the war was in progress; the existence of a body dedicated to predict the future which had now become the present seemed superfluous.

Increased Funds and Opportunities

Let us go back to the period of the Ethiopian war. It has already been mentioned that the increase of CNR funds in 1935 was followed by an even greater augmentation in 1936, which raised its disposable financial resources to 10 million liras (more than 8 million euros). The increase was apparent before the Second World War and funding eventually reached more than 25 million liras (almost 17 million euros) per annum. Thus, within five years, the real value of CNR funds was multiplied by more than 17 (Maiocchi, 2000 and 2003, p. 264). Mussolini's initiative was fundamental for this great increase. This sudden wealth brought new perspectives to the CNR. It was the first time that putting the original programme into practice, at least partially, became possible. If nothing else, the unit was able – also thanks to the contribution of many companies which Mussolini asked to intervene – to build its own headquarters, which were opened in 1937.

The increase of funds was not accompanied by any governmental directive regarding scientific research that should be considered overriding for Italian autonomy. In other words, Mussolini gave much money, though he did not say how to spend it. The CNR was forced to invent a role for itself that it would play in the process of the construction of imperial Italy. This often provoked the jealousy of various ministries, above all of the Ministry of Education, whose head was Giuseppe Bottai.

In early 1936, when the victory of the Italian army in Africa seemed imminent, the CNR, similarly to many Italian public bodies, launched a project to contribute to the evaluation of the resources that could be found in the lands that were being conquered. The Council wanted to demonstrate through its own diligence that it was worth the funding increase granted by the Duce. It proposed, therefore, to cooperate with ministries and bodies but the answers it received, however, were disappointing (Maiocchi, 2003, p. 115). So, the formation of a commission for mining prospects in the colonial lands had been proposed to the Ministry of Colonies, but no sign of approval had arrived from it. Therefore, such a proposal was addressed directly to Mussolini. It intended to organise a mission that, following the Italian heroic soldiers' advance, would carry out mining prospects on the conquered lands and would 'search for auriferous sand in the Tigrai'. The Duce did not answer, while the Ministry of Colonies finally informed the Council that it was not interested in the proposal of the CNR. In January 1936, the Academy of Italy had agreed to send scientific missions to Ethiopia, so the management of the CNR asked the Academy to be allowed to collaborate in sounding out the Ethiopian riches. The Academy did not answer until the end of the war operations, in May, when it communicated that it did not mean to carry out mining research in the colonies. The CNR repeated its proposal of a mining commission to the Ministry of Colonies, which this time gave a vague answer. Since minerals did not seem to generate much success, the CNR proposed a photogrammetric survey of Ethiopia, but this, too, went nowhere.

The only thing the CNR managed to achieve was the organisation of a commission of chemists, which in 1936 explored Ethiopia in search of industrial structures that later might be further developed. After its return to Italy, the commission painted a depressing picture with no interesting perspective, and so the final report was absolutely useless. It is worth underlining the fact that the head of the Ethiopian mission was Henry Molinari, a recognised expert on plant design and installation who, however, was well-known also by the Italian police as a militant anarchist. Because of his political ideas, Molinari was forced to quit university and could not obtain a permit to leave the country. It was only due to a personal intervention by Mussolini that Molinari was given a passport so that he could leave for Africa (Maiocchi, 2003, p. 122). Also, in the following years Molinari occupied important posts in the CNR. It seems that Mussolini accorded more importance to technical competences than to political fidelity. As for Molinari, not once did he show – in public speeches – that, regardless of his political anti-Fascist position, he was captivated by the idea of autarky. In his view, from the perspective of scientific research the autarky project was the most rational solution. This is only one example of the approach that characterised many Italian technicians: that the autarky-project, interpreted as an evaluation plan of the national resources by means of scientific research, seemed an absolutely reasonable idea.

Although it could not provide any relevant contribution to the war in Africa, the CNR still had the possibility of realising ambitious research programmes, thanks to its increasing funds. However, the idea of forming national laboratories had to face the obstacle of the regulatory framework, according to which the CNR was not free to found its own institutes, being still under the authority of the Ministry of Public Education, which in the meantime had been re-named the Ministry of the National Education. Marconi and Giannini asked Mussolini to release the CNR from the Ministry. In 1936, the head of the Ministry was Giuseppe Bottai, a high-ranking political figure of the Regime. Bottai had no intention of letting the CNR out of its sphere of influence, and therefore he declared himself decidedly against Marconi's proposal. They both tried to convince the Duce about the validity of their positions, in a dispute that ended with Marconi's victory. In June 1937, a decree with the force of law was published, reorganising the CNR and giving it the authority to build its own scientific laboratories.

Nevertheless, this victory cost a lot. The clash with the Ministry had caused an almost total paralysis of the CNR: the management held no meetings for almost a year, from 27 July 1936 to 28 June 1937, while the secretary took care of the ordinary business. Resumption of activity had to be postponed again, because in the following July Marconi died, and it became necessary to designate a new president. The position was offered to Pietro Badoglio, who was appointed in November 1937. The new management met for the first time in December 1937. Actually, the CNR remained inactive, except for ordinary business and the continuation of old research projects, until the end of 1937. This period was an extremely important year and a half for Italy, which ended the war in Africa and began the policy of autarky, in view of a war on a larger scale. The great amount of money the CNR had received could not be spent, so the management decided to buy State bonds. The appointment of Badoglio, who was the Chief of Staff (the highest military authority after the King and Mussolini) can be explained only if we bear in mind the climate of mobilisation in Italy. In summer 1937, Badoglio, who had led the army in the successful expedition to Africa, was extremely popular and also extremely wealthy. Probably, by offering him the charge (with a pay of about 100,000 liras per year, about 75,000 euros), Mussolini meant to take advantage of his prestige in order to build around the CNR a network of relationships that would go beyond scientific circles, also stressing the possible role of the institution in the Nation's preparation to war. Badoglio moved his military offices to the new centre of the CNR, thus creating a physical link between the army and the scientists. However, Mussolini's expectations were disappointed, and the CNR was never held in great consideration by the army.

There were many signs of detachment between the CNR and the military sphere. For example, in June 1938, the Supreme Commission for Defence (*Commissione Suprema di Difesa* – CSD) discussed the possibility of substituting iron, copper, tin and antimony with other materials that Italy possessed in greater abundance. Badoglio (for the first and last time) tried to involve the CNR and remarked that this kind of evaluation could have been done by the institution chaired by him. The other members of the CSD did not grant the CNR much credibility, since they asked for a 'formal assurance' that the CNR would be really capable of dealing with such a problem. Badoglio addressed the question to Pio Calletti, the president of the Committee for engineering of the CNR. Calletti could not do anything more than forming a commission to answer the question of whether the CNR would be able to tackle the problem. At the same time, Calletti asked Badoglio for additional funds for this new commission. Badoglio angrily answered that it was unacceptable to form a commission for every matter; he dropped the project and later avoided offering any aid between the CNR and CSD.

In the years 1938–39, the CNR started to work at full power. Their funding was subsequently increased. According to the official declarations, the CNR would have to direct all of its forces towards autarky, but in reality things followed a different pathway (Maiocchi, 2001a). First of all, a decree stated that the CNR was to use a large part of the funds at its disposal to constitute a national geophysical service and to reconstruct the National Thalassographic Committee – two institutions which were not linked to the autarky-project. Moreover, no political or military body continued to instruct the CNR about the strategies to be followed when formulating scientific research projects of autarkic interest. Mussolini said nothing more, no ministry asked for assistance – on the contrary, the animosities of the preceding years continued and the Armed Forces, regardless of Badoglio's presence, did not seem to regard it as useful to involve the CNR in their own activities. Thus, once again, the Institute had itself to invent a role to play. The management of the CNR, however, was formed mostly of people whose background was not scientific and who did not have qualifications (as had explicitly been recognised) to formulate plans regarding Italian scientific research. Therefore, since nobody created any plans regarding autarky-orientated research, no one ever indicated the priorities needed on the endless list of problems brought into discussion every day by the autarky-construction issue.

Everything remained entrusted to the initiative of individuals who managed to obtain funding for their own studies due to their personal contacts rather than because of the objective importance of their research questions. Many of those researchers who now appeared as autarky-constructors put forward the same issues that they had already dealt with in the preceding years and that had previously not gained attention, but which now became extremely fashionable in the new ambience of autarky (Maiocchi, 2003, chapter 3). These researchers represented the 'scientific–technical nationalism' which had appeared during the Great War: to them autarky meant the realisation of an ideal they had pursued for a long time without success. Among names that could be enumerated here the most significant is that of Mario Giacomo Levi. Levi, lecturer at the Technical University of Milan, for almost two decades had studied the features and the possible uses of Italian coal as an alternative to the imported anthracites. With the appearance of autarky, Italy's shortage of coal seemed to be the fundamental problem of economic production and Levi's studies suddenly became famous. In a speech delivered in the autumn of 1937 on the change that came about Levi said:

> In 1931, at the 20th meeting of our Society in Milan I was to speak about a part of the problem, that is, about the technical and economic aspects of the fuel issue. My faith, my enthusiasm and our work did not slacken ... but the atmosphere in Italy was sceptical and fearful: what prevailed were strictly economic considerations ... I admit that I suffered during this Congress. I left the meeting discouraged and bothered by doubts about whether it was true that I was obsessed and fanatical about my insistence upon studying problems which to our Country meant neither possibility, nor benefit ... And how different is the atmosphere today! ... The land cultivated with conscientious faith germinates vigorously, the indifferent have become enthusiastic, the incompetent rushed to study and have become scholars, the industrialists, the technicians, the capitalists are fully mobilized, our 130 publications are being searched, read and sold everywhere. The reasons for such a change are known to everybody: for the third time in twenty years the problem of fuel has recently reappeared in Italy, displaying all its violent gravity – maybe more violent than ever because the whole World has united or has tried to unite against us, when 50,000 Italians were abroad in another continent, conquering the Empire. A brilliant victory or suffocation and humiliation depended on transport, production and weapons; the only really national and really available raw material [is] the heroism of our soldiers of all units and in all ranks, the prophetic clairvoyance and the super-human courage of the Duce (Levi, 1937, p. 297).

Yet, in autumn 1938 Levi was expelled from the University and was persecuted under racial laws.

Just like Levi, many other scientists offered their scientific credibility in favour of autarky, even when the latter became a plan of preparing Italy for an exceptionally important war. Public support of the scientists involved in the autarky project was of great propaganda value and served to add a touch of 'being scientific' to programmes which were all but reasonable.

The CNR and the Second World War

With Italy entering the war in May 1940, there were many – and many useless – offers of help to the army that came from different institutes of the CNR. A good example is the case of the Central Commission for the Examination of the Inventions of the CNR (since 1941 it was named the National Institute for the Examination of the Inventions of the CNR). This commission had already begun its activity in 1939, in order to create a Special Permanent Committee with the task of examining the proposals for war machines coming from public institutes, industries or even private citizens, and then tasked with informing the army immediately about the proposals that were most worthy of consideration. The Committee worked intensely and pointed out the inventions that seemed to be applicable to war operations. There was, however, never any sign of interest from the military authorities, despite the solicitations and the complaints of Tito Montefinale, the Commission's president.

There were just a few occasions on which the CNR had a chance to contribute to the war effort. It seems that the most significant activities were those of the Inter-ministerial Commission for substitutive matters, which I have already mentioned, and those of the Experimental Centre for the Applications of Psychology, which was created in Rome in March 1940 by Agostino Gemelli and which was directed by Ferruccio Banissoni. The Centre worked on the selection of military personnel. Other relevant institutions were: the Engine National Institute of Naples, directed by Pericle Ferretti, where important research on turbines for underwater navigation was carried out and eight torpedoes for aviation were built; the Experimental Centre of Torre Chiaruccia, which carried out research on the application of short waves to ground links, and last the National Institute for the Applications of Calculation, which answered many calculation problems given to it by the military authorities.

Badoglio's presence at the CNR did not provide any link with the army, and his activity as president was quite insignificant: he seldom made interventions in debates, where direction was left to general secretary Frascherelli, and he almost never made remarks about scientific–technical questions. His contribution was limited to bookkeeping observations about the operations' economic feedback, and to sour comments about the limited usefulness of many studies.

Starting from Italy entering the war, the CNR was gradually abandoned to its destiny. Neither Mussolini, nor any other member of the government, nor any ministry, nor any army institution seriously asked the CNR to collaborate for the war, leaving it without any indication of its utility. Its funds were considerably diminished by the inflation caused by the war. A great part of its finances was spent on obeying two orders that had nothing to do with the war needs: the constitution of the National Geophysical Service and the re-organisation of the Committee for Marine Geography, which included a number of institutes that studied the physics and the biology of the sea and inner waters. The remaining money was spent on incoherent research projects proposed by single researchers. Sometimes their aim was the national interest, but at other times their inspiration was just personal fancy. Of course, the project of building great national laboratories for physics, chemistry and biology was put aside. The frightening situation in which Italy found itself during the German invasion, after 8 September 1943, did not spare the CNR, whose activity became feebler and feebler until it almost ceased completely. After the war, the CNR would be born again on completely new grounds.

Conclusion

I shall conclude this chapter with a brief overview of the research conducted in the political–institutional climate I have sketched out. The research produced in this context was of various levels and produced diverse results. First of all I should certainly mention research which could be conducted only because of autarky and which led to failure. The group that is usually referred to in order to describe the particular scientific climate of this period must be divided into two sub-groups. One group is constituted by typically Italian research topics, such as work on substitute textile fibres (Lanital, 'cottonized' hemp) or the use of plants like broom as sources of cellulose, while the second group consists of research which, due to the technologies applied, were eventually to be forgotten but which, in a given moment, could be considered as in line with the international scientific community: such were the studies of gasogene material, to which the CNR dedicated its largest research institute, the Engine Institute (*Istituto Motori*) in Naples. Reference models for this kind of research were France, Germany, Switzerland, Austria. Reference models also included research on reinforced concrete with bamboo cane instead of iron (along the lines of what was being done in Germany), conducted with great intensity in the centre for studies on construction material in Turin led by Gustavo Colonnetti, who at the same time was working also on an avant-garde issue, namely pre-compressed concrete. Still, along with the efforts which could only be justified by the climate of the time (and they were not limited to Italy only), which were doomed to be instantly forgotten and to which one used to emblematically reduce the whole science of the second half of 1930's, other typologies were also present. Research lines which had already been followed autonomously in the past were resumed by scholars who finally found a way to make their names known and became the centre of general attention in the autarkic climate. This renewed certain areas of research, for example studies regarding the use of national combustibles, the production of aluminium and light alloys, the extraction of cellulose from annuals. Also new research projects, stimulated and made possible by the autarkic conjuncture, were initiated. These studies, which would later be significantly developed, included Giulio Natta's research on the production of synthetic rubber supported by IRI and Pirelli. Also this type of research constituted a prelude to Montecatini's achievements in the field of plastic material in the post-war period, as well as to Natta's personal success in the field of polymerisation. There was also industrial research based upon foreign patents, without the contribution of the university circles, which gave birth to great advances in production such as the hydrogenation of combustibles by Anic or the production of national magnesium in Bolzano. Also, again without the contribution of the university, original industrial research took place which brought important results, such as the perfecting of the T4 explosive by Nobel.

This mobilisation, more operational than ideological, of scientists and technicians was not and could never have become sufficient to give any plausibility to the autarky project. Our shortages of raw material and of production capacity were too large, too disastrous, to achieve the ambitions of autarky, even in such a limited and partial shape as was sketched in the fascist plans.

List of References

Levi, M.G., 1937. 'Autarchia dei combustibili', *Atti Riunione SIPS*, 1937.

Maiocchi, R., 1993. 'Scienziati italiani e scienza nazionale *(1919–1939)*'. In: S. Soldani and G. Turi (eds), 1993. *Fare gli Italiani. Scuola e cultura nell'Italia contemporanea.* Vol. II. Bologna: Il Mulino, II vol., pp. 41–86.

Maiocchi, R., 1998. 'Gli istituti di ricerca scientifica in Italia durante il fascismo'. In: R. Simili (ed.), 1998. *Ricerca e istituzioni scientifiche in Italia.* Roma-Bari: Laterza, pp. 182–212.

Maiocchi, R., 2000. 'L'organizzazione degli scienziati italiani'. In: V. Calì, G. Corni and G. Ferrandi (eds), 2000. *Gli intellettuali e la Grande guerra.* Bologna: Il Mulino, pp. 209–44.

Maiocchi, R., 2000a. 'La ricerca scientifica nel CNR'. In: A. Casella, A. Ferraresi, G. Giuliani and E. Signori (eds), 2000. *Una modernità difficile. Tradizioni di ricerca e comunità scientifiche in Italia 1890–1940*. Pavia: La Goliardica, pp. 405–30.

Maiocchi, R., 2000b. 'Le istituzioni di ricerca in Italia tra le due guerre'. *Atti della Fondazione Giorgio Ronchi*, 55, pp. 691–714.

Maiocchi, R., 2000c. 'Fascism and Italian Science Policy'. In: L. Guzzetti (ed.), 2000. *Science and Power: the Historical Foundations of Research Policies in Europe*. Brussels: European Commission, pp. 179–86.

Maiocchi, R., 2001. 'La ricerca agraria'. In: R. Simili and G. Paoloni (eds), 2001. *Per una storia del Consiglio Nazionale delle Ricerche*. Roma-Bari: Laterza.

Maiocchi, R., 2001a. 'Il CNR da Badoglio a Giordani'. In: R. Simili and G. Paoloni (eds), 2001. *Per una storia del Consiglio Nazionale delle Ricerche*. Roma-Bari: Laterza, pp. 173–200.

Maiocchi, R., 2003. *Gli scienziati del Duce. Il ruolo dei ricercatori e del CNR nella politica autarchica del fascismo*. Roma: Carocci.

Maiocchi, R., 2004. *Scienza e fascismo*, Roma: Carocci.

Simili, R. and Paoloni, G. (eds), 2001. *Per una storia del Consiglio Nazionale delle Ricerche*. Roma-Bari: Laterza.

Chapter 8

Statistical Theory, Scientific Rivalry and War Politics in Fascist Italy (1939–1943)[1]

Jean-Guy Prévost

Corrado Gini's 1939 lecture on *The Dangers of Statistics* is considered as a landmark in the history of Italian statistics. Delivered at the outset of the Italian Society of Statistics' (*Società Italiana di Statistica* – SIS) first meeting, it became the launching salvo of a full-blown assault against the inferential methods characteristic of 'Anglo-Saxon' statistics. Over the next four years, Gini would devote no less than a dozen papers to an elaborate 'critical review of the foundation of statistics'. Such an effort on the part of Italy's then-foremost statistician may be described first and foremost as a reaction to recent developments that had profoundly transformed the theory and practice of statistics as a transnational discipline. For instance, while the 1925 International Statistical Institute (ISI) report on 'the representative method' had maintained an agnostic position between 'random' and 'purposive' selection, Jerzy Neyman's 1934 paper on stratified random sampling clearly established the superiority of probabilistic methods; and Ronald Fisher's 1925 *Statistical Methods for Research Workers* had become the gospel of practical statistical research in many domains, leading to the increasing use of significance tests as a 'mechanized inference process' (Gigerenzer, 1987, p. 18). At the same time, the fact that Gini's critical review was conducted during the Second World War, in the context of an authoritarian–totalitarian regime and along lines that paralleled those of the ongoing conflict, surely deserves serious consideration.

Our intent here is therefore to move away from the kind of interpretation that envisions the history of science as the rational reconstruction of concepts and techniques to a more historically and sociologically grounded analysis. In our view, the concept of statistics as an intellectual or scientific 'field', which may be defined as a structured and multidimensional set of positions governed by specific criteria of legitimacy, offers a more appropriate framework for taking

---

[1] The author wishes to thank Daniela Cocchi for her comments and suggestions.

into account how a discipline's autonomy is always intertwined with, sometimes enhanced and at other times hindered by, the evolving environment in which it operates (Bourdieu, 1997; Prévost, 2009). The institutional – by contrast with the narrowly scientific or conceptual – development of Italian statistics, personal, domestic and transnational rivalries among statisticians, the specific historical and political circumstances, together provide overlapping contexts against which to interpret the meaning of Gini's 1939–43 contributions. From this combination of perspectives, Gini's criticisms of mainstream inferential methods cannot be isolated from his more general scientific ambitions and from his considerable experience as a 'practical' statistician. Besides being a statistician, Gini was a potent scientific entrepreneur, an unorthodox economic theoretician, as well as an idiosyncratic intellectual of the Fascist era. All these identities did fashion, to one degree or another, the views about probability and inference he put forward from 1939 on. Internal analyses of Gini's intellectual 'war effort' have generally framed the issues at stake according to rival views regarding probability. By taking into account the various forms of rivalry that were at play, one may argue that Gini's intentions were much more overarching: no less than an attempt to provide 'an alternative to classical statistical inference' (Piccinato, 2011, p. 102), a new statistical orthodoxy for a new world order in which Anglo-Saxon hegemony over statistics and science in general would have perished as a correlate of the Allies' defeat on the battlefield.

The Disciplinary Perspective: Internal and Retrospective Interpretation

Taken together, Gini's wartime papers on inference have been described as 'an attack on Fisher's fiduciary methods and the Neyman-Pearson theory of hypothesis testing' and an argument for 'the importance of prior probabilities in judging the measures of a sample' (Brooks, 2001). They directly challenged the basic assumptions behind significance tests and the theory of confidence intervals. In his 1939 lecture, Gini asked, for instance, if the error of a given constant having a probability P_a of being accidental, one should conclude that it also had probability $1 - P_a$ of being significant (Gini, 2001, p. 271).[2] In contemporary parlance, the error Gini pointed at was 'the fallacy of the transposed conditional', that is, drawing from the inconsistency of an observation with the null hypothesis an argument in favour of the hypothesis

---

[2]   A number of Gini's papers have been collected and translated in Gini, 2001. We use that translation when possible.

one champions. In a later paper, in reference to quotations drawn from Fisher's *Statistical Methods* and Yule and Kendall's 1937 *Introduction to the Theory of Statistics*, Gini called attention to 'the misunderstanding between the probability that an observed result depends on accidental causes and the probability that the said result is obtained when only accidental causes intervene' (2001, p. 310). He also cast doubt upon the general validity of the hypothesis of 'random error compensation', to which he opposed 'the prevalence principle of constant causes' and underlined the difference between the probability of a given value and the probability that such a value resulted from chance rather than from systematic causes (Gini, 1941a).

Gini's 1939–43 papers have been interpreted by later Italian statisticians in the light of contemporary theoretical debates about probability and inference and of Gini's own early contributions in this regard (Forcina, 1987; Costantino, 1994; Giorgi, 2005). These authors have generally characterised Gini's views as 'prepar(ing) the ground in Italy for the re-birth of Bayesian theories' (Herzel and Leti, 1978, p. 26), as 'empirical Bayesian' (Forcina, 1982, p. 65) and, more recently, as 'strictly objective Bayesian' (Frosini, 2005, p. 435). In this sense, a parallel may be drawn between Gini's criticisms and the position defended by Harold Jeffreys during his dispute with Fisher in the early 1930s and in his 1939 *Theory of Probability*. But in the late 1930s, Bayesian methods, prior probabilities and inverse probability had reached a low point in the opinion of most statisticians. The Fisher and Neyman-Pearson approaches, which were based on a frequentist (probability as 'a limiting ratio in a sequence of repeatable events') rather than epistemic or Bayesian conception of probability (probability as 'a degree of knowledge or belief' making use of prior information) had become dominant. They also had a ring of scientific objectivity that appealed to a number of social scientists in search of legitimacy (Howie, 2002, pp. 187–98). The fact that the more radical version of Bayesianism, which re-emerged in the 1950s, can be traced back to the 1930s writings of Bruno de Finetti – a mathematician who had worked with Gini and a founding member of the SIS – may explain the attention given in retrospect to Gini's wartime writings on inference. A recent account has described this episode as 'a missed opportunity' and characterised Gini's criticisms as 'destructive', by comparison for instance with De Finetti's 'constructive' attitude (Piccinato, 2011, pp. 112–13); besides, it is fitting that, given the context, Piccinato's account uses words such as 'violent attack' and 'destructive' to characterise Gini's attitude.

Yet, as commendable and useful the above-mentioned interpretations may be, they take for granted a degree of autonomy that practitioners of the discipline have fought hard to develop and maintain. It is for instance obvious that the definition of a statistician in Italy and the scope of the discipline have evolved significantly between the period under consideration here (1939–43) and that when statisticians of a later generation published their assessments of Gini's views. These changes had largely to do with the progressive demarcation between, on the one hand, statistical methodology and formalised models and, on the other hand, the empirical subject matters to which they could be applied (Frosini, 1989, pp. 207–10). The comprehensive view of statistics that Gini and many of his generation entertained has faded away on the aftermath of the war, in part because of the mathematisation of the discipline and of the increased specialisation it induced. It is also clear that, in the late 1930s, the 'internationalisation' of statistics had not yet been achieved. It may seem awkward to speak nowadays of 'Italian' statistics, but such an originality was acknowledged for a considerable time, as shown by the presence of an appendix on Italian methodological contributions in four editions (1937, 1940, 1944, 1947) of Yule and Kendall's *Introduction to the Theory of Statistics* and, even more significantly, by the inclusion of more than sixty asterisked entries prepared by Gini himself in the first four editions of the authoritative International Statistical Institute (ISI) *Dictionary of Statistical Terms* (1957, 1960, 1976, 1982). At the same time, the 'hybrid theory' that combined key concepts from Fisher and Neyman-Pearson while conflating differences between their respective frameworks and to which many practitioners and experimenters soon rallied was still in the making during the 1930s (Gigerenzer, Swijtink, Porter, Daston, Beatty and Kruger, 1987, pp. 106–9). Should one therefore interpret specific episodes in terms of their embodied results in the present state of a discipline or in terms of a still undefined future and thus opened to more than one path? As we shall see, moving beyond the purely theoretical content of scientific arguments and taking into account the various rivalries – personal, institutional and international – in which these arguments were embedded and played out provides us with a much richer picture and understanding of the various issues at stake.

Personal Rivalry: the Gini–Fisher–Neyman Triangle

By 1939, Corrado Gini (b. 1884) was a towering figure in Italian statistics. He had secured his reputation thanks to a well-received dissertation on sex ratio at birth and a path-breaking series of papers on probability written between 1907 and 1911. Up to then, Italian statisticians were basically interested in compiling demographic, economic and social data; in contrast with mathematicians and engineers, most had scant knowledge of probability calculus. In these papers, Gini spelled out a number of issues that were central to probability – with distinctions between 'classical' and 'empirical' conceptions of probability, 'deductive' and 'inductive' applications, 'a priori' and 'a posteriori' probabilities – and established the importance of their practical consequences for all those who were engaged in the analysis of numerical series. After a span at the University of Cagliari (1907–13), Gini was awarded the prestigious Padua chair of statistics in 1914. That same year, he published the memoir in which he defined R, nowadays known as the Gini concentration coefficient, and thereby ensured his place in the pantheon of Statistics. When Italy entered the war, Gini was entrusted with various duties regarding the organisation of the war effort; after the armistice, he likewise brought his expertise to a number of commissions, conferences and official bodies concerned with reconstruction. In 1925, at a time when many intellectuals began to entertain second thoughts and put their name to Benedetto Croce's *Manifesto of Anti-Fascist Intellectuals*, he rather chose to support Giovanni Gentile's rival *Manifesto of Fascist Intellectuals*. In 1926, Gini was chosen by Mussolini to oversee the reorganisation of official statistics and become the head of the Central Institute of Statistics (*Istituto Centrale di Statistica* – ISTAT) and of its advisory body, the High Council of Statistics (*Consiglio Superiore di Statistica* – CSS). Gini took up his functions with dedication and enthusiasm, but he kept his university chair (now in Rome), which allowed for an impressive concentration of power and prestige. Even though he was constrained to resign from ISTAT and the CSS in 1932 following bureaucratic infighting, Gini could still rely on considerable resources throughout the 1930s, having command over an unrivalled network of journals, research centres and disciples and playing a decisive role in academic appointments. During the 1920s and 1930s, his impressive editorial output dealt mainly with descriptive statistics, demography and economics, but Gini's comprehensive view of statistics and social science led him to claim authority also over research in sociology, ethnology, eugenics and genetics. A very active member of the ISI, Gini was frequently invited abroad and, on numerous occasions, he stressed the contrast between the 'Italian School of Statistics', focused on the description of frequency distributions, with special attention to averages, variability and concentration, and 'Anglo-Saxon' statistics, mostly concerned with sampling, statistical significance and the management of error.

Ronald Aylmer Fisher (b. 1890), for his part, embodied the latter kind of statistics, which was to become the mainstream in the English-speaking world, thanks notably to the huge success of his 1925 *Statistical Methods for Research Workers* and his 1935 *Design of Experiments*. Fisher was an altogether different figure than Gini and he clearly outranks him in the pantheon. He was thoroughly trained in mathematics, whereas Gini was largely self-educated in that subject. Fisher was also less dispersed than Gini in his pursuits: besides his achievements in mathematical statistics, he is remembered today as the father of population genetics and a decisive architect of the modern evolutionary synthesis in biology. By 1939, Fisher was, like Gini, a highly respected figure in statistics, but his influence extended beyond his country of birth, with a core of disciples in the United States and in India. Fisher's relations with Gini went back to the early 1920s, when a paper of his, critical of Bayesianism as well as of Karl Pearson's interpretation of inverse probability, was first rejected by *Biometrika* and then published in *Metron*, the new journal founded, managed and owned by Gini. Three other papers by Fisher were published in *Metron*, in 1924 and 1925. The two statisticians also met in person at least three times: in 1924 and 1928, at two successive International Congresses of Mathematics held respectively in Toronto and Bologna, and in 1936, on the occasion of a Cowles Commission conference in Colorado. From at least 1931, Gini and Fisher wrote regularly to each other. The tone of their letters was polite and friendly and they show Gini keeping up with Fisher's writings throughout the 1930s. Fisher made specific comments on Gini's October 1939 lecture in three letters dated 22 December 1939, 19 January 1940 and 3 May 1940, which Gini quoted in a 1943 lecture on significance tests. In these, Fisher conceded that if his or some of his colleagues' 'language' regarding statistical significance may have been at times 'arrogant' or 'peremptory', their 'practice' and their 'applications' were basically sound. Gini was not satisfied with this concession 'in the matter of logic', however, and he was especially worried by Fisher's insistence that tests could be applied 'without setting up any detailed hypotheses as to the possible causes to which any apparent disturbances might be due' (Gini, 2001, pp. 212–13, 312–15). Communication ceased after Italy entered the war in June 1940 and resumed only in 1945. In a letter dated 4 July 1945, Gini promised to send reprints of his wartime papers on inference, but there is no trace of any further exchange on the subject.[3]

Polish mathematician Jerzy Neyman's relations with Gini were more difficult from the start. They first met in 1928, on the occasion of the Bologna congress, where Gini was one of those responsible for section IV, on 'statistics, probability calculus, mathematical economics and actuarial sciences'. Neyman, who was still pretty green (he had received his PhD in 1924 at the age of 30), presented a strongly anti-Bayesian paper on the methods for estimating hypotheses. In the general session, he enthusiastically pushed forward the idea that all contributions on probability, which were dispersed amid a wide range of journals, be published in

---

3   Correspondence between Gini and Fisher has been archived by the University of Adelaide Library. See: http://digital.library.adelaide.edu.au/dspace/handle/2440/67701.

one or very few outlets, so as to enable researchers to keep track of developments. He went on to suggest, without having previously consulted with interested parties, that the *Rendiconti del Circolo Matematico di Palermo* or *Metron* play that role. After discussion, Neyman's original motion gave way to the more realistic proposal of setting up a bulletin summarising all papers devoted to probability that were published in various languages (Unione Matematica Internazionale, 1929, pp. 105, 119, 126–7). This event may be regarded as a significant date in the institutional history of probability as a branch of mathematics and it surely testifies to the prestige of Italian mathematics at that time (De Finetti, 2006, p. 4). But 'young' Neyman's initiative may have appeared presumptuous to Gini (who, aware of his limited mathematical credentials, cautiously insisted on differences between mathematicians and statisticians). The two met again in 1929, at an ISI meeting in Warsaw; Gini, now in a more familiar setting, 'vigorously attacked' another paper Neyman presented on hypothesis testing (Reid, 1982, p. 86).

But the ground on which Neyman and Gini would collide was sampling. As a government statistician, Gini was concerned with the practical issues of sampling, and since Norwegian statistician Anders Kiaer's experiments on 'representative enumerations' at the end of the nineteenth century, the topic was discussed both inside and outside of the ISI. In 1925, an ISI commission to which Gini belonged delivered its report on the 'representative method' (Jensen, 1925). It presented two methods as equally valid. One was random selection, defined as that according to which units of a sample should be chosen through a process that ensured all units of the population have an equal chance of being selected. The other was purposive selection, according to which a sample was valid when its units presented characteristics similar to those of the universe from which they were extracted; representativeness of such a sample could be measured for instance by comparing its dispersion with that of the population. Gini's name became closely associated with this method following his attempt, with the help of Luigi Galvani, at constructing a purposive sample of the 1921 Italian census forms, threatened with destruction for want of space. Gini and Galvani were intent on saving a subset of these in order to conduct further analysis of the census results. To this end, they chose as their sample 29 of the 216 census districts, on the basis that, in the case of seven major variables, averages did not differ much from those of Italy as a whole. Gini and Galvani soon discovered, however, that when they looked at other variables or at values other than the mean for the same variables, these districts were not representative anymore. This led them to draw a radical distinction between 'relative representativeness', which occurred when sample and population coincided in one or more aspects or variables, and 'absolute representativeness', which they deemed impossible (Galvani and Gini, 1929, pp. 22–5).

In 1934, when Neyman addressed the Royal Statistical Society on the respective merits of stratified random sampling and purposive selection, he devoted a large section of his paper to Gini and Galvani's experiment. Even though Neyman's paper was a thorough demolition of the Italian statisticians' work, its tone remained quite civil. For instance, he reminded his public that 'the Italian statisticians (...) did not find their results to be satisfactory' and summarised the problem they had faced in the form of a paradox: 'the consistency of the estimate suggested by Gini and Galvani, based on a purposely selected sample, depends upon hypotheses which it is impossible to test except by an extensive inquiry' (1934, pp. 585–6). In a lecture delivered at the Graduate School of the United States Department of Agriculture in 1937, Neyman was however much more caustic. In the meantime, he had, with Egon Pearson, developed the method of confidence intervals and acquired an enviable reputation. Now, he described purposive selection as 'most dangerous and (...) practically certain to lead to deplorable results'; what Gini and Galvani provided above all was 'a good example of how not to sample human populations' (1952, pp. 104–5). Of course, Gini was aware of these remarks, since he mentioned both Neyman lectures in his 1943 paper on significance tests.

The point here is not of course that jealousy or pettiness alone can account for theoretical disagreements. Gini, Fisher and Neyman were three difficult personalities and there was no love lost between Gini and Neyman (nor between Fisher and Neyman). But personal rivalry was here the embodiment of two neatly different overall conceptions of statistics. Both Neyman and Fisher were mathematicians and they envisioned statistics as 'essentially a branch of applied mathematics' (Fisher, 1925, p. 1), while Gini's motto was 'statistics with the least mathematics possible' (Gini, 1926, p. 706). These self-definitions nicely capture two possible futures for statistics: one in which models, formalisation, mathematics and probability become the core of the discipline against one in which connection with empirical subject matters remains much closer. Confidence in sampling had developed with the appeal of frequentist probability, of which Neyman had become one of the staunchest defenders, while the epistemic view, which raised the issue of prior knowledge, induced towards scepticism. Gini tended to distrust sampling as a cheap and satisfactory substitute to exhaustive and costly inquiries as did other 'practical' statisticians like Arthur L. Bowley, also a member of the 1925 ISI Commission (but Bowley completely changed his views after hearing Neyman). Such a view was not atypical among government statisticians. Exhaustive inquiries remained the reference; the national census, with 'many individuals but few variables', was the grand undertaking and the monographic method, with 'few individuals but many variables', remained popular, notably in Italy (Desrosières, 2008, p. 110). Given 'the primitive nature of the data at our disposal' and the fact that it 'may be too rough to allow of the application of exquisite methods', Gini thought it 'more useful' to spend time and money 'increasing the number, sometimes too limited, of observations' (1926, pp. 706–7). By the late 1930s, American pioneers of survey sampling, such as Morris Hansen and W.E. Deming, who had been instrumental in bringing Neyman to the United States, argued that not only was sampling less costly, but it could yield more accurate results, as far as control over all phases of the survey was improved. But even after the war, Gini still promoted 'accuracy' and 'completeness' of data as an ideal and presented 'the indiscriminate use of sampling' as a dangerous fashion (2001, pp. 424–5).

Institutional Rivalry: Two Societies for a Single Science

Gini's decision to launch a sustained critique of the inferential methods championed by Fisher and Neyman may also be set against changes in the field of Italian statistics in the late 1930s. The quasi-simultaneous creation, in late 1938 and early 1939, of two competing statistical societies, with distinctive membership and orientation, was the awkward outcome of an inconclusive debate held in 1935 about the wisdom of creating an Italian society of statistics. It was also the reflection of a partial overhaul with regard to the power and resources individuals or groups could command, following Gini's resignation from ISTAT and the CSS. Up to then, statisticians and economists had been loosely gathered in sections 2A (actuarial mathematics, mathematical statistics, probability calculus) and 14 (economic and social sciences) of the Italian Society for the Progress of Sciences (*Società Italiana per il Progresso delle Scienze* –SIPS). Meetings of the SIPS were held nearly every year since its rebirth in 1907; but no national body of statisticians on the model of Britain's Royal Statistical Society or of the American Statistical Association existed. Interestingly, scepticism about the desirability and feasibility of such a grouping was clearly expressed by Gini himself. Other statisticians were also sceptical and had expressed fear that such a body may be subject to 'monopoly' or 'domination' by certain persons, groups or schools – a transparent allusion to Gini's position and character. In early 1937, taking advantage of the split that had occurred between the International Union for the Scientific Study of Population (IUSSP) and Gini's own Italian Committee for the Study of Population's Problems (*Comitato italiano per lo studio dei problem della popolazione* – CISP) following a clash over the scientific independence of the 1931 Rome International Congress on Population under the latter's chairmanship, Florence-based statistician and demographer Livio Livi launched the Advisory Committee for the Studies on Population (*Comitato di consulenza per gli studi sulla popolazione* – CCSP), which immediately became Italy's representative in the IUSSP.

The CCSP nucleus was composed of some thirty statisticians, demographers and economists, none of them close to Gini. Its scientific legitimacy was unquestionable, since nine of Italy's seventeen full professors of statistics, seven members of the CSS, as well as the two top ISTAT officers, President Francesco Savorgnan and Director General Alessandro Molinari, joined its ranks. In November 1938, the CCSP changed its name to Italian Society of Demography and Statistics (*Società Italiana di Demografia e Statistica* – SIDS). Membership quadrupled, with eminent or politically well-connected figures, such as Rodolfo Benini, dean of Italian statisticians, or Alberto De Stefani, member of the Fascist Grand Council and former Finance minister (1922–25). In its statutes, the SIDS was dedicated to demographic and statistical studies alike. Its main promoter, Livi, commanded undeniable scientific eminence as a demographer; at the head of various journals and demographic or economic research units, he had emerged as a powerful academic entrepreneur and a contender to Gini's position in the field. Population policy was of course a central issue in Fascist Italy and the SIDS would move close to the Regime's priorities, with Livi and Savorgnan joining the High Council for Demography and Race (*Consiglio Superiore della demografia e della razza*), an advisory body created following the enactment of racial laws. As head of ISTAT, Savorgnan was also to be found among the ten signatories of the July 1938 *Manifesto degli scienziati razzisti*, published as a rationale for the anti-Jewish legislation.[4] In 1939, links with the Regime's most unsavoury policies became more evident, when Livi claimed that the creation of the SIDS had been 'supported with sympathy' by the Department of Demography and Race (*Direzione generale della demografia e della razza*), the main body in charge of applying racial policies and that it had been 'entrusted' with conducting 'specific enquiries' on behalf of *Demorazza*, as it became known (Società Italiana di Demografia e Statistica, 1940, p. 11).

The Italian Society of Statistics (SIS) was for its part created in January 1939, two months after the SIDS, as a response from statisticians close to Gini. Its

---

[4]   For a detailed analysis of this episode, see Maiocchi, 1999, pp. 212–40. By a tragic irony, the SIDS, which, contrary to the SIS, was directly involved in demographic and racial policy work, saw many of its founding members become victims of the racial laws.

original nucleus gathered about the same number of academic statisticians, but very few economists. Mathematical statisticians were much more present in the SIS than in the SIDS and among them was Bruno de Finetti, the rising star in probability theory. Interestingly enough, Gini did not appear as one of the SIS's original promoters; all the more spectacular was his entering the stage as the keynote speaker of its inaugural meeting in October 1939. In a regime where organic unity was the proclaimed norm, especially after the totalitarian turn of the late 1930s, the creation of a second statistical society obviously needed a rationale. This was provided by the insistence, in article 1 of the SIS's statutes, on the development of 'scientific research in the field of statistical disciplines with special attention to statistical methodology' (Pietra, 1939). By comparison, the SIDS's statutes, and its own article 1, insisted on 'the progress of demographic and statistical studies, with special attention to the quantitative and qualitative progress of the Italian population' (Società Italiana di Demografia e Statistica, 1939). The demarcation was therefore clearly spelled out: the SIS was dedicated to *scientific research*, by contrast with the SIDS's more vaguely worded *studies* and the connection to policy this implied; statistics was the single overarching object of the SIS and methodology, rather than an empirically-grounded concept such as population, provided its specific focus.

Moreover, notwithstanding the loyalty oath to the King and the Fascist Regime article 7 imposed upon its president and vice-president, the SIS remained at arm's length of official bodies like ISTAT or *Demorazza*. In his earlier intervention on the opportunity of creating a statistical society (1935), Gini, no more at ISTAT, had declared such a degree of scientific independence to be necessary: otherwise, he argued, its members would not have felt free to criticise the statistical production of official bodies. Meetings of the SIDS (eight were held between 1938 and 1943) were often concerned with policy issues, such as labour, insurance statistics or the economy and demography of the Mediterranean area. By contrast, Gini's highly theoretical inaugural lecture set the course for the SIS, where more than a third of all papers (55 out of 145) presented on the occasion of the seven meetings held between 1939 and 1943 dealt with methodological topics. At the same time, SIS meetings were the place where the architectonic definition of statistics, that is, statistics as a set of methods but capable of making significant contributions to empirical subject matters, was reasserted, as shown by the diversity of contributions also presented on these occasions (Società Italiana di Statistica, 1964). In other words, for Gini and his followers, statistics had a disciplinary existence of its own, but social sciences were subordinated to the statistical method, a point Livi had challenged with regard to demography, which he considered a fully fledged science (1941, pp. 2–3). Besides reasserting the singularity of the Italian outlook *vis-à-vis* mainstream Anglo-Saxon statistics, Gini's critique of inference also acted as a clear demarcation line in a field where eminence, prestige and access to resources were in constant flux. Bringing back probability as a central and foundational issue for a correct assessment of statistical methods and directly challenging the emerging orthodoxy on that issue was not a task the SIDS as a body, nor most of its members, were up to.

International Rivalry: Scientific Nationalism and Gini's 'Second Front'

As mentioned earlier, Gini was eager to insist on the originality of Italian contributions to statistical methodology. He had done so in a lecture before the London School of Economics in 1926 and he came back on the subject in a paper posthumously published in the *Journal of the Royal Statistical Society* in 1965. In both cases, he provided exhaustive assessments of the Italian statisticians' achievements. To be sure, the accomplishments of the 'Italian School' according to Gini were but a selective and skewed subset of the work done by the community of Italian statisticians. This was also the case with the Yule and Kendall appendix, where, of the 78 titles, no less than 48 bore Gini's signature, while seven were by Gaetano Pietra, four each by Francesco Paolo Cantelli and by Galvani, all close to Gini; the remaining titles were divided among 14 authors mentioned once or twice. And, of course, the ISI dictionary entries dealt exclusively with work carried out by Gini, who was the sole Italian taking part in this project. Divisions and rivalry within the Italian statistical field, which led to the creation of two statistical societies, were distortedly mirrored at the international level.

Gini's proselytising was in line with what Roberto Maiocchi has coined as 'technical-scientific nationalism', that is 'an ideology ... which, by means of conferences, publications, organisational initiatives and political pressure, made an attempt to obtain concentration of resources ... and a greater involvement of Italian scientists in research of applied character' (2009). Now, by the late 1930s, Italy was subject to economic sanctions following its invasion of Ethiopia and scientific nationalism was on the rise. The meeting held on the occasion of the SIPS's centennial in October 1939 – two days after the SIS's foundational meeting and some forty days after Germany's invasion of Poland – provided indeed a platform for celebrating Italy's scientific achievements. During the session on statistics, Gini and statisticians close to him delivered an overall assessment of the state of the discipline, dealing with probability calculus, methodology, biometry, demography and economic statistics. In all five areas, the reports chauvinistically reviewed early Italian contributions and work accomplished since the turn of the century. They all insisted on Italy's intellectual and scientific independence in a manner that was consonant with the normative constraints of 'autarky', now a centrepiece of the Regime's official discourse. Gini's own report, which sought to establish the 'international position' of Italian statistics, was written from a comparative standpoint. It envisioned the issue almost as a sporting contest, with Italy and England standing *ex aequo* with regard to methodology (albeit mostly – and thus somewhat ungentlemanly – because of the English – imperial – language being disseminated all around the world); in demography, Italy clearly led the way, while England kept the upper edge in biometry; there was clear American supremacy in the field of economic statistics – yet, only from a 'quantitative point of view', since, 'in the narrowly scientific domain, the contribution of Italian statistics (was) not inferior to that of any other nation' (1939, pp. 245–52). Gini's string of papers criticising developments made by 'the English School' to the theory and practice of inference nicely dovetails the contrast on which this assessment was built.

This opposition between national schools also coincides, in the most obvious sense, with the alignment of powers in the war that had just started. In fact, the context of the war can be connected directly to the views Gini expressed in his 1939–43 contributions on probability and inference. First, it is obvious that Gini did not have a complete grasp of the positions he criticised and that he 'did not appreciate the difference between the Fisher and Neyman-Pearson approaches' (Forcina, 1982, p. 67). In a 1945–46 paper recapitulative of his wartime production, Gini imputed this to difficulties of communication during the war, notably to the fact that he could not take into account a paper Neyman had published in 1941 (2001, p. 380). However, Forcina shows that Gini's errors betray insufficient familiarity with or misunderstanding of earlier papers by Neyman and Fisher (1982, p. 67). Obviously, the war could not but accentuate scientific 'autarky' and opposing the English and Italian schools had the effect of conflating differences between Fisher and Neyman as well as of muting those among Italian statisticians. The fact that the two statisticians invited at the 1943 SIS meeting to defend the value of significance tests, M.P. Geppert and H. Von Schelling, were Germans also illustrates limitations to cross-national scientific exchange; Gini was able to dismiss easily these opponents and the problem was therefore not 'discussed with full competence' (Forcina, 1982, p. 67).

Finally, given the fact that Gini's 'critical review of the foundation of statistics' represented a considerable investment of energy as well as a significant part of his wartime output (twelve contributions between October 1939 and June

1943), it is interesting to compare it with the rest of his intellectual production during that period.

Indeed, other topics sustained Gini's attention between the outbreak of the Second World War and the Fascist Regime's fall in July 1943, but three groupings stand out. Eight papers, most of them delivered at SIS meetings, dealt with various aspects of statistical methodology (such as transvariation, correlation or interpolation). A dozen presented results from Gini's bio-demographical research and were mostly published in his journal *Genus*. But Gini's most consistent set of papers during the war (11) dealt precisely with the global conflict and its various aspects. Among the topics covered were the population policies of democracies and totalitarian countries (five papers), the importance of access to raw materials in a world of contending powers (two), the optimal size of post-war economic blocks with regard to autarkic development (three) and opinion polls as an instrument of political manipulation (one). These interventions were saturated with the Regime's ideological conventions, with positive references to 'autarky', titles such as 'the crisis of the bourgeoisie and the duty of totalitarian regimes' and calls for *Grossvolkplanung* (large-scale demographic planning); 'democracies' were presented as the enemy, fortunately driven to decay by powerful internal tendencies. But these papers also put into play the intellectual resources of statistics. In his 1941 paper on 'the struggle between conservative and expansionist peoples', for instance, Gini argued for the impossibility of democratic decision-making by making use of the Condorcet paradox (1941b, pp. 410–11). In his 1942 interventions on the definition and possibility of autarky, he tested six different configurations of a Mediterranean economic zone by comparing to what degree they could be self-sufficient with regard to 18 basic commodities (Gini, 1942, pp. 221–58). Some of the outlets in which these papers featured were official or semi-official journals of an overtly ideological character such as *Razza e civiltà* or *Archivio di Studi Corporativi*. Two lectures were delivered in Berlin in February 1942, a time when many were still confident in Germany's victory.

Taken together, these papers envisioned a New World Order dominated by the victorious German–Italian axis and may be considered, besides those on inference, as Gini's 'second front'. The parallel is in fact striking between the two series of writings, which, together, make up Gini's overall intellectual contribution to the Nazi–Fascist war effort, as if the 1939–43 papers directed against Anglo-Saxon inferential methods were the theoretical sublimation of the 1940–43 papers concerned with the more practical issues of population and economic policies. As a coda to all this, we may mention that Gini later wrote that the intellectual victory of the Anglo-Saxon school and the 'indiscriminate' and 'blind' use of 'mathematics' by 'a large number of young statisticians' in Italy resulted from the outcome of the war (2001, pp. 422–3). In other words, as the British used to say that 'trade follows the flag', for Gini, it was also Science that followed the flag!

Conclusion

Gini's wartime criticism of 'Anglo-Saxon' statistical inferential methods stands as a significant episode of his own intellectual path as well as of the history of Italian statistics. It also stands as an enlightening example of scientific theoretical activity in the context of a totalitarian regime at war. From the perspective of statistics as a discipline, it can be interpreted as a 'contribution to the logical bases of inference' and as a one-man struggle against the then-dominant trend (Herzel and Leti, 1978, p. 6). But, from the wider perspective of statistics as a field, it also offers lessons as to the relations between scientific theory and various layers of the social and political environment in which it is produced. Notably, this episode allows us to understand in what sense we can assert the existence of a structural homology – as distinguished from analogy – between the content of scientific theoretical activity (in the present case, the calling into question of an interpretation of probability and inference) and the social and political context in which it is thought out, embraced and defended.

A first level of analysis is that of statistics as a field of institutions and positions, with its own struggles for power, prestige and resources. In this context, Gini's contributions on probability and inference provided a rationale for the creation of the SIS as well as a clear demarcation with the rival SIDS. Significantly, this occurred at a time when contenders seemed to threaten Gini's long-standing hegemony over the field. In other words, to the theoretical divide between Gini's views on probability and inference and those he criticised corresponded an organisational division between opposing groups of Italian statisticians. This alignment was not, to be sure, a direct reflection of the theoretical divide about inference, in the sense that the choice made by statisticians as to which statistical society they joined had nothing to do with the frequentism/inverse probability issue. But the result was portrayed as a divide between, on one side, theory, methodology and scientific autonomy and, on the other, policy-relevant applications. In the contest between the two rival societies, Gini's SIS portrayed itself from the start as the most 'scientific', a trait that would ensure its lasting success, even though Gini's views on inference eventually faded away.

ITALIAN STATISTICAL FIELD

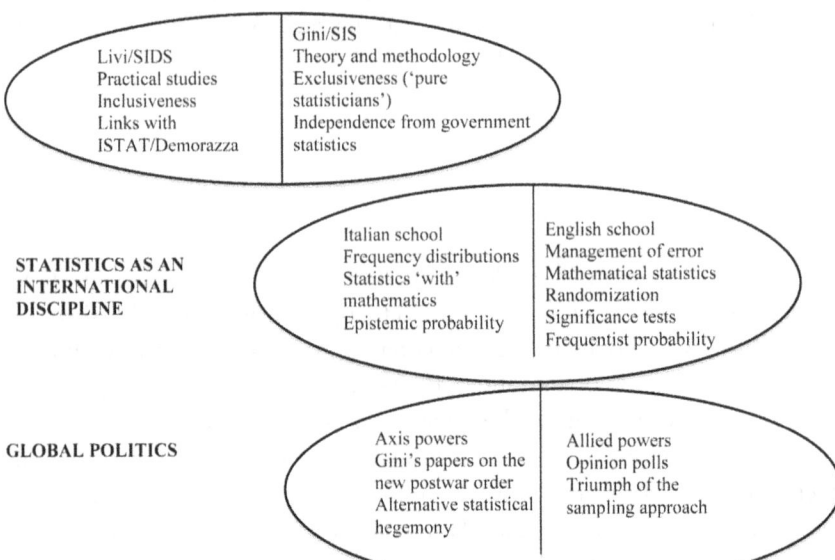

Figure 8.1   A three-level structural homology

A second level is that of statistics as an international discipline. Appearing as a major player on this level – as shown by Gini's presence at ISI meetings – could not but reinforce one's position on the domestic scene. But Italian politics, notably the nationalistic character of the Regime, also had a clear impact. It is for instance obvious that methodological nationalism was very much alive in the 1930s and that it acted as a prism that could both magnify and obscure various aspects. Differences between Italian and Anglo-Saxon statistics were undoubtedly real and important, but it is clear that a number of high-profile internal disputes occurred on the Anglo-Saxon side (between Fisher and Jeffreys, between Neyman and Fisher), which were highly relevant to Gini's points and yet largely ignored by him. The intellectual autarky that resulted from the Regime's normative constraints seems to have been here more important than communication breakdown. The chronological coincidence between the creation of a statistical society primarily devoted to methodological research and the launching of a string of contributions that would provide it with a singular 'line of attack' both on the national and the international statistical scene is also significant.

But not only was the SIS more 'theoretical', and therefore more comprehensive in its view of statistics, than the SIDS, it was also more 'Italian'. First, there was a considerable overlap between the SIS and 'the Italian statistical school' as admittedly recognised by authoritative authors of the English tradition. Secondly, it was the SIS that raised the flag against 'the enemy'. This brings us to a final and higher level: rivalry among nations in the paroxysmal war context. Here, politics came back with a vengeance and completely over-determined the theoretical-methodological divide. Opinion polls and the sampling approach, both of which were rebuffed by Gini, played a major part in managing support for war and the war effort on the Allies' side, especially in the United States. And what appeared, in the domestic context of rival statistical societies or in that of international statistics, as a purely theoretical issue became part of a two-pronged intellectual war effort. While Gini resorted, on the battlefront of probability, to the abstract and euphemistic language of theory, at the same time, he devoted all the resources of statistics as a practical knowledge to the cause of Nazi–Fascist post-war domination over territory, population and resources.

List of References

Bourdieu, P., 1997. *Les usages sociaux de la science. Pour une sociologie clinique du champ scientifique*. Paris: INRA.

Brooks, B., 2001. Tales of Statisticians. Corrado Gini. [online] Available at: <http://www.umass.edu/wsp/ statistics/tales/gini.html> [Accessed 16 January 2012].

Costantino, D., 1994. 'Corrado Gini sulle orme di Bayes et Laplace'. In: *Le scienze matematiche nel Veneto dell'Ottocento*. Venice: Istituto veneto di scienze, lettere ed arti, pp. 263–72.

De Finetti, F., 2006. Some links between Bruno De Finetti and the University of Bologna. [pdf] Available at: <http://www.brunodefinetti.it/Bibliografia/ Bruno De Finetti and the University of Bologna.pdf> [Accessed 26 May 2013].

Desrosières, A., 2008. *Pour une sociologie historique de la quantification*. Paris: Mines Tech.

Fisher, R.A., 1925. *Statistical Methods for Research Workers*. London: Oliver and Boyd.

Forcina, A., 1982. 'Gini's Contributions to the Theory of Inference'. *International Statistical Review*, 50(10), pp. 65–70.

Forcina, A., 1987. 'Gini's View on Probability and Induction'. In: A. Naddeo (ed.). *Italian Contributions to the Methodology of Statistics*. Padua: CLEUP/ Società Italiana di Statistica, pp. 327–34.

Frosini, B.V., 1989. 'La statistica metodologica nei convegni della SIS'. In: Società Italiana di Statistica (ed.). *Atti del convegno 'Statistica e Società'*. Pisa: Società Italiana di Statistica, pp. 197–227.

Frosini, B.V., 2005. 'Objective Bayesian intervals: some remarks on Gini's approach'. *Metron*, 53(3), pp. 435–50.

Galvani, L. and Gini, C., 1929. 'Di una applicazione del metodo rappresentativo all'ultimo censimento italiano della popolazione [1° dicembre 1921]'. *Annali di Statistica*, series 6, no. 4.

Gigerenzer, G., 1987. 'Probabilistic Thinking and the Fight against Subjectivity'. In: K. Lorenz et al. *The Probabilistic Revolution*, vol. 2. Cambridge: MIT Press, pp. 11–34.

Gigerenzer, G., Swijtink, Z., Porter, T., Daston, L., Beatty, J. and Kruger, L., 1987. *The Empire of Chance*. Cambridge: Cambridge University Press.

Gini, C., 1926. 'The Contributions of Italy to Modern Statistical Methods'. *The Journal of the Royal Statistical Society*, 89(4), pp. 703–24.

Gini, C., 1935. 'Per una Società italiana di statistica'. *Barometro economico italiano*, 72, pp. 295–6.

Gini, C., 1939. 'Introduzione; caratteristiche e posizione internazionale della statistica italiana'. In: Società Italiana per il Progresso delle Scienze. *Un secolo di progresso scientifico italiano*, vol. 1. Rome: Società Italiana per il Progresso delle Scienze, pp. 245–52.

Gini, C., 1941a. 'Alle basi del metodo statistico. Il principio della compensazione degli errori accidentali e la legge dei grandi numeri'. *Metron*, 14(2–4), pp. 173–240.

Gini, C., 1941b. 'La lotta attuale tra popoli conservatori e popoli espansionisti e l'evoluzione organica delle Nazioni'. *Archivio di studi corporativi*, 12(3), pp. 379–456.

Gini, C., 1942. 'Autarchia e complessi economici supernazionali'. In: *Convegno per lo studio dei problemi economici dell'Ordine Nuovo, Pisa, 18–23 maggio, Atti, Vol. I, Relazioni*. Pisa: Arti grafiche Pacini Mariotti, pp. 221–58.

Gini, C., 2001. *Statistica e induzione/Induction and Statistics*. Bologna: CLUEB.

Giorgi, G.M., 2005. 'Gini's scientific work: an evergreen'. *Metron*, 63(3), pp. 299–315.

Herzel, A. and Leti, G., 1978. 'Italian contributions to statistical inference'. *Metron*, 35(1–2), pp. 3–48.

Howie, D., 2002. *Interpreting Probability. Controversies and Developments in the Early Twentieth Century*. Cambridge: Cambridge University Press.

Jensen, A., 1925. 'Report on the Representative Method in Statistics'. *Bulletin de l'Institut international de statistique*, 1, pp. 359–80.

Livi, L., 1941. *Trattato di demografia – Le leggi naturali della popolazione*. Padua: CEDAM.

Maiocchi, R., 1999. *Scienza italiana e razzismo fascista*. Florence: La Nuova Italia.

Maiocchi, R., 2009. 'Fascist Autarky and the Italian Scientists'. *Journal of the History of Science and Technology*, 3. [online] Available at: <http://johost.eu /?oid=89&act=&area=5&ri=2&itid=3> [Accessed 4 April 2012].

Neyman, J., 1934. 'On the Two Different Aspects of the Representative Method: The Method of Stratified Sampling and the Method of Purposive Selection'. *Journal of the Royal Statistical Society*, 97(4), pp. 558–625.

Neyman, J., 1952. *Lectures and Conferences on Mathematical Statistics and Probability*. Washington: Graduate School of U.S. Department of Agriculture.

Piccinato, L., 2011. 'Gini's criticisms to the theory of inference: a missed opportunity'. *Metron*, 69(1), pp. 101–17.

Pietra, G., 1939. 'La Società italiana di statistica'. *Supplemento statistico*, 4(4), pp. 1–7.

Prévost, J.-G., 2009. *A Total Science. Statistics in Liberal and Fascist Italy*. Montreal/Kingston: McGill-Queen's University Press.

Reid, C., 1982. *Neyman from Life*. New York: Springer Verlag.

Società Italiana di Demografia e Statistica, 1939. *Atti della IIIa Riunione, Bologna, 13 novembre 1938–XVII*. Florence.

Società Italiana di Demografia e Statistica, 1940. *Atti della IVa Riunione, Roma, 27–28 Maggio 1939–XVIII*. Florence.

Società Italiana di Statistica, 1964. *Sommario e Indice per autore degli atti delle riunioni scientifiche della Società italiana di statistica 1939–1963*. Rome: Fausto Failli.

Unione Matematica Internazionale, 1929. *Atti del Congresso Internazionale dei Matematici*. Bologna: Zanichelli, vol. I.

Chapter 9

Science, Military Dictatorships and Constitutional Governments in Argentina[1]

Pablo Miguel Jacovkis

Introduction

Argentina has a long tradition of conflicts between scientific development[2] (or educational development) and authoritarian governments.[3] Between 1835 and 1852, although he was only Governor of the Buenos Aires Province and responsible for Argentinian Foreign Affairs – that is, a sort of *primus inter pares* amongst the other thirteen governors – General Juan Manuel de Rosas ruled the country with an iron fist. Among his obscurantist measures we may mention that he stopped paying salaries to the professors at the University of Buenos Aires – the salaries having to be paid exclusively through the tuition of the students, see Buchbinder (2005). The simplified image is reasonably true, that from the establishment of the constitutional republic between 1852 and 1862, education and support for science flourished. We can mention the outstanding work of President Sarmiento (1868–74), who, besides strongly backing

[1]   The author thanks Amparo Gómez for her kind invitation to participate in the *IV Seminar of Politics of Science: Science between Democracy and Dictatorship* at the Universidad de La Laguna and to the Spanish Ministry of Science and Innovation for its economic support through the Research Project FFI2009–09483 and the Complementary Action FFI2010–11969–E. He also thanks Rosita Wachenchauzer and Israel Lotersztain for their comments and observations, and also Amparo Gómez and Antonio Fco. Canales and, especially, Brian Balmer, for their careful review and criticisms of this work, although of course none of them is responsible for the opinions here expressed.

[2]   In this work we shall focus on natural and exact sciences, without mentioning social sciences and humanities.

[3]   Anyway, in Argentina the relationship between authoritarian (in the twentieth century, usually military or with a strong military influence) governments, on the one hand, and science and technology, on the other hand, was ambiguous. The military appreciated that science is needed to develop industry and the technical potential of the nation, but they distrusted Argentinian scientists. We shall describe this ambiguity in this work.

elementary education, created, during his administration the (later named) National Meteorological Service, the National Academy of Sciences in Córdoba and, also in Córdoba, the Astronomic Observatory. In particular, the creation of the Observatory is quite symbolic, as it was first necessary to have an ambitious idea of what science means in order to create an astronomic observatory in a country of two million inhabitants which had just ended a terrible war, in which 80 per cent of the population was illiterate, then locate it in a small city of thirty thousand inhabitants, and also convince an internationally known astronomer – Benjamin Gould – to be in charge.[4] Sarmiento's approach to science has other very remarkable characteristics, such as his lecture in homage to Darwin in 1882, in which he demonstrated a stunning knowledge of science. Sarmiento was not alone; the so-called 'generation of the eighties' encouraged secular elementary education. Juan María Gutiérrez, as Rector of the University of Buenos Aires, founded the Department of Exact Sciences, which eventually became the Faculties of Exact and Natural Sciences, of Engineering and of Architecture, Design and Urbanism.[5] Following this line of thought, we may also mention the creation of the National University of La Plata in 1905, the support its Rector Joaquín B. González gave to science and many other similar developments.

There were also shadows: neither Gutiérrez nor González were successful in their efforts to create genuinely scientific universities. The difficulties, for example, that the future Nobel Prize winner Bernardo Houssay had to overcome to be appointed Professor of Physiology at the Faculty of Medicine of the University of Buenos Aires show that the prevailing culture at that University was not to hold scientific activity in high regard (Vaccarezza, 1981). There were even outbreaks of authoritarianism which provoked, for instance, a select group of secondary education teachers to create a private high school after resigning from their positions at the *Colegio Nacional de Buenos Aires*, the flagship of secondary education for upper class children (Sanguinetti, 2006). But, overall,

---

[4]   Gould remained in Argentina for more than 13 years where he developed an amazing amount of scientific activity, especially recording stars from the southern hemisphere. He also directed the National Meteorological Bureau (the future National Meteorological Service). A glimpse of his professional activity in the USA may be seen in Galison (2005). For a more detailed analysis of Gould's life, see Bernaola (2001).

[5]   The Department was created in 1865. As Halperin Donghi (1962) describes, Gutiérrez entrusted the Italian physician Paolo Mantegazza with the hiring of well-paid professors in Italy. Three Italian professors were hired in this way. It is remarkable that, in spite of the huge economic – and political – problems due to the war against Paraguay, that (for Argentina) began the same year, the authorities maintained the project of appointing the professors. It is clear that the hiring was part of a scientifically – or, at least, academically – ambitious project.

the 'liberal country' that existed between 1862 and 1930, in its elitist version until 1916 and in its mass-democratic version between 1916 and 1930, either supported scientific development or did not obstruct it.[6] And during this period, in 1918, University Reform began in Córdoba, which among other things, attempted to modernise the archaic university structures. It spread like wild-fire, not only across all the country but also across Latin America. The Reformists enthusiastically supported science, although Bernardo Houssay, who as mentioned was the most distinguished scientist contemporary with the University Reform, and several of his colleagues, opposed it.[7]

The Military Incursion into National Politics

From 1930 onwards the situation became more confused. That year, a military *coup d'état* led by José Félix Uriburu, a general with fascist tendencies, overthrew President Hipólito Yrigoyen, and the long period of constitutional governments, which had begun in 1862, was interrupted. Uriburu's quasi-fascist project was unsuccessful: reluctantly, he was forced to call a general election and a new period of constitutional governments began. This new period was based on this first election, which outlawed the 'yrigoyenists' and, in addition, the period was marked by electoral fraud.[8]

The first President of this new period was also a professional military man, Agustín P. Justo (1932–38), who had collaborated with the military coup. At the same time, the Catholic Church recovered part of the influence

---

[6]   In fact, the second option, that the ruling class did not obstruct scientific development, is closer to the truth. This is in the sense that the ruling class did not actively harm it: except for a minority of members of the elite, including firstly the already mentioned people (and particularly Sarmiento, the politician and statesman who most supported science in the nineteenth century, and perhaps most in all our history), the ruling class of that time showed more indifference towards science than actual support for it.

[7]   The relatively modest success of the Argentinian reformists in transforming the universities into scientific centres of excellence requires a detailed analysis beyond the scope of this work. Hurtado (2010) carefully describes the (also failed) attempts of Houssay and his colleagues, around the 1940s, to create private universities with a scientific orientation in order to solve the problem of low scientific activity in public universities.

[8]   The triumphant group, among the sectors who had supported Uriburu's *coup d'état*, favoured the elitist republic from before 1916; this regression was impossible without electoral fraud.

it had lost during the elitist republic.[9] Justo's administration was a right-wing administration, whose legitimacy was questioned by many citizens. Significantly, Justo does not fit the traditional image that the intellectuals adopted some years later of uncouth and ignorant coup-prone military men: Justo was an engineer besides being a military man and was quite cultivated. Perhaps an indication of his and his collaborators' intellectual capacity is the skill with which Argentina managed to reduce the worst effects of the 1930 world crisis (faster than many other countries).[10]

General Enrique Mosconi was another very capable and learned man and also an engineer, who was fired from his position as Director of YPF[11] as soon as Uriburu took power, and would never have thought of participating in a plot against a constitutional government. On the other hand, the 'father' of the Argentinian iron and steel industry, General Manuel Savio, was a right-wing nationalist who participated in Uriburu's coup. In a sense, the Argentinian iron and steel industry was born 'in a state of sin', because of Savio's influence. Savio was representative of the strong nationalist right-wing group in the Argentinian Armed Forces who, during the period (1930–83) in which the Armed Forces exerted an immense power in Argentina, be it directly or by pressure on civilian governments, could never solve a crucial problem:[12] for those nationalists, the

---

[9]   Given the low legitimacy of the new administration, it needed the Catholic Church's support, which in the elitist republic had not been necessary. The relationship between the Church and General Justo's administration (and the subsequent administrations until 1943) is described in Zanatta (1996), who also describes the process of 're-Christianization' of the Argentinian Army, which until then had been rather liberal (in the nineteenth-century meaning of this word).

[10]   For instance, Hernández Andreu (1987) points out that 'the Argentinian growth rate was higher than the Canadian and American ones during the 1930s'.

[11]   *Yacimientos Petrolíferos Fiscales* (YPF) was a State-owned firm that competed against the powerful foreign oil companies. Mosconi transformed YPF in the 1920s into an important company. Until its privatisation in the 1990s YPF symbolised Argentinian nationalism in opposition to the big international oil firms, for the left as well as for the nationalist right.

[12]   The power of the Armed Forces, which reached its zenith during the last dictatorship (1976–83), began to dissolve after the defeat in the Falkland (Malvinas) war against Great Britain (1982), which forced them to call a general election. The winner was Raúl Alfonsín who, as soon as he was inaugurated as President, ordered that the members of the military *Juntas* during the dictatorship be judged. With ups and downs, the military's power began to decrease significantly, until it practically disappeared during Carlos Menem's and Néstor Kirchner's administrations. In fact, symbolically the disappearance of military power in Argentina may be represented by the replacement of all members of the general staff on 25 May 2003, when Kirchner, elected with a scarce 20 per cent of the votes, was inaugurated.

Armed Forces should be powerful, so that Argentina could be in a position to face Brazil (and/or Chile), according to the theories of conflict studied at the time.[13] In order to have powerful Armed Forces, Argentina should be technologically developed and industrialised. Technological development was obviously related to scientific development. The unsolvable problem for the military was that many scientists (not all, of course) were 'dangerous leftists' or, at least, did not trust the Armed Forces.[14]

So, on the one hand, some sectors of the Armed Forces wanted to incorporate scientifically and technologically trained people to their industrialist project and, on the other hand, suspected many of these people of communism or something similar. Unlike Mosconi, who until he died in 1940 was not particularly appreciated by the governments originating in the 1930 coup, Savio got along very comfortably, not only during the fraudulent conservative governments that survived until the following *coup d'état* in 1943, but also during the 1943–46 military government; the latter was originally much more authoritarian and right-wing than the previous one, and within which some of its members embraced a strong fascist ideology. Indeed, except for the fact that the situation was tragic, the subsequent repression of university students and faculty by the military authorities, including the imprisonment of many of them, would seem to be an exaggerated caricature of military–clerical hatred of science.[15] Recall that shortly after coming to power, the military authorities changed the name of the National Secondary School of Buenos Aires (*Colegio Nacional de Buenos Aires*) to Saint Charles University School (inspired by the name the school had during colonial times) and appointed Juan Sepich, a priest, as principal (Sanguinetti, 2006).[16]

Furthermore, during a ceremony, before the eyes of the President, the Army Chief of Staff removed the portraits of former military dictators Jorge Rafael Videla and Reynaldo Bignone on 24 March 2004.

[13]   Nationalist military men were extremely concerned about Argentinian military power in the context of a military government (1943–46) and the Second World War (in which Argentina was neutral until the last moment possible). This concern is reflected in the fact that, in 1945, 43.3 per cent of the national budget was set aside for the Armed Forces (Potash, 1981).

[14]   This contradiction refers to mathematics and natural sciences, and their corresponding technologies. In this work we shall not analyse how military men considered social sciences and humanities.

[15]   As already mentioned, from 1930 onwards the Catholic Church began to recover the influence it had held until the liberalism of the 1880s generation; part of the (successful) strategy of the Church in this regard was to increase its influence amongst the Armed Forces.

[16]   This change of name was short-lived.

It is important to draw a line between the relationship of the Armed Forces with science and technology before and after 1930, that is, before and after their irruption as a protagonist of Argentinian politics until 1983. There was not a sudden change: on the one hand, after 1930 the Armed Forces followed – or tried to follow – policies that were not too different from prior ones; on the other hand, their process of ideologisation, involving a special mysticism and 're-Christianization', had begun before 1930.[17] What happened in 1930 was a qualitative and quantitative change to the influence of the military in national politics. In previous years, its not inconsiderable influence was expressed in the traditional way of constitutional democratic countries (for example, the opinions of important officers, articles in periodicals and newspapers, and so on). From 1930 on, the military either had the right of veto or overthrew governments which did not accept their right of veto. They began to have a power and a responsibility that they had not had before, and their relationship with technologists and scientists, and their interest in science and technology, became a direct part of national policies.

Armed Forces, Science and Technology

The interest of the military in science and technology existed from Argentina's beginnings as an independent country. In fact, after the May 1810 Revolution and the replacement of the Spanish Viceroy with a new government not appointed by Spain (the First *Junta*), in September, 1810, the *Junta* created a School of Mathematics and Manuel Belgrano, one of its members, delivered a significant speech, in which he said that a young officer would find in this institution '... all the tools provided by mathematical science applied to the lethal, although necessary, art of war' (Babini, 1986). Anyway, the period until the definitive constitution of the Argentinian State in 1862 was very unstable, so that it is worth observing that, for the Armed Forces, it was only from this year that any discussion about national projects, tacit or explicit, was possible.

The period 1850–1950 is described very well by Ortiz (1992): originally, the military and civilian higher education were not divided, but they increasingly separated, due, for instance, to the creation of the Technical Higher School of

---

[17]   Perhaps one can establish, as the beginning of this process, the speech delivered in Lima by the great poet (turned into a right-wing nationalist) Leopoldo Lugones in December 1924, for the centenary of a battle which secured independence from Spain of its South American colonies. Here, he uttered the unfortunate sentence: 'The hour of the sword, for the good of the world, has again arrived' (Lugones, 1979).

the Army. Some important military men of great intellectual capacity (Justo, Mosconi) had become engineers studying at the universities, but eventually the officers obtained their technical training in military institutions, and this phenomenon contributed to the separation of military men from civilians. As their influence on Argentinian politics increased after 1930, many of them began to think of themselves as a kind of aristocratic elite and this feeling permeated their relationship with science and technology. It cannot be ruled out that this trend had been combined with the German influence in the Army (much less in the Navy, which had been influenced by Britain) and with the influence of the Catholic Church, so as to create a contradictory situation between the pro-industry wishes of the military and the mistrust *vis-à-vis* scientists and technologists.

The Perón Era

General Juan Domingo Perón was elected President in 1946 in the first non-fraudulent elections since the 1930 *coup d'état*. His administration was marked by two conflicting aspects: persecution and support. On the one hand, there were persecutions, firing of faculty and scientists, repression (including his unfortunate role in a scientific–technological fiasco, in which he trusted a pseudo-scientific trickster who claimed to be able to perform nuclear fusion under controlled laboratory conditions, which turned out to be an international faux-pas after spending a lot of public money).[18] On the other hand, the period saw the creation of the National Agency for Atomic Energy (*Comisión Nacional de Energía Atómica* – CONEA) and the development of the aeronautics industry. Both in the nuclear fusion fiasco and in support for the aeronautics industry, German scientists and technologists who had emigrated to Argentina after the Second World War participated in meaningful ways: in the nuclear fusion affair the protagonist was the Austrian Ronald Richter and in the aeronautics project Kurt Tank[19] (who had influenced in the decision of bringing Richter to Argentina). Tank developed – jointly with other émigré German engineers – the Pulqui II airplane, of which five prototypes were made, the last of them in 1959. In any event, there is a fundamental difference between Tank and Richter:

---

[18]   The story of this fiasco is told pleasantly and carefully in Mariscotti (1985). It is very interesting to see that Perón paid more attention to German scientists and engineers than to the Argentinian scientific community.

[19]   The arrival of Tank and some of his collaborators has been described for instance in Goñi (1998).

there can be no doubt regarding Tank's professional quality, regardless of any discussion on the real value of the Pulqui II models.[20] It is possible to strongly criticise the huge budget spent on Project Pulqui II but, on the other hand, that kind of undertaking potentially had additional benefits that were difficult to take into account but not negligible: training of personnel, suppliers with professional expertise that outlived their main client, spill-over of state-of-the-art technology, and so on.[21] It is worth mentioning that, before Pulqui II, there was a Pulqui I project, which also took place during Perón's administration, but without much success. One of the main participants was the French engineer Emile Dewoitine, who fled France due to the fact that he collaborated actively with the Vichy Régime.[22]

Peronism Outlawed

Although some distinguished professors maintained their positions at the universities, in spite of political repression and discrimination against professors who did not join the Peronist Party, many intellectuals, scientists and professionals, leftists and rightists, suffered the interruption of their academic careers due to persecution.[23] So, it is not surprising that, after Perón's fall in 1955, due to a military *coup d'état* with civilian support in an extremely polarised society, they returned enthusiastically to the universities. Many of them contributed to the university revival between 1955 and 1966, which was considered by many people as the 'golden age' of Argentinian universities.[24] In

---

[20]   A detailed history and discussion of Project Pulqui II may be consulted in Artopoulos (2007).

[21]   Anyway, those projects in the military area almost always failed in Argentina. In particular, Pulqui II, as Artopoulos (2007) describes.

[22]   In 1948 Dewoitine was sentenced (in absentia) to twenty years of hard labour (Klich, 1999).

[23]   Not only the persecutions. For many professionals and intellectuals the authoritarianism of the government, the pressure on people to become members of the government party, and what could be called (in current terminology) 'cult of personality' (for example, the name of the President and of his late wife given to streets, cities, provinces; the manuals for elementary school which praised the authorities) were a sufficient reason either to distance themselves from the government or to participate in the opposition to it.

[24]   Buchbinder (2005) offers a general idea of the evolution of universities in that period; Halperin Donghi (1962) analyses the University of Buenos Aires, but his account, although very detailed, does not cover the final phase of this period, subsequent to the writing of his book. As a particular example of progress in that period in one discipline, computer science, see Jacovkis (2006) and Factorovich and Jacovkis (2009).

fact, historians and journalists could remark on the curious phenomenon of a military coup in Argentina which, unlike all previous (and subsequent) coups, supported democratisation of the universities and scientific and technological development.[25] Incidentally, those concepts do not necessarily go together: Houssay's attitude *vis-à-vis* a democratic university is a clear counterexample.[26] Moreover, during the military government (which outlawed Peronism and lasted until 1958, when constitutional President Frondizi was elected) the National Institute of Farming Technology (*Instituto Nacional de Tecnología Agropecuaria* – INTA), the National Institute of Industrial Technology (*Instituto Nacional de Tecnología Industrial* – INTI) and the National Council of Scientific and Technical Research (*Consejo Nacional de Investigaciones Científicas y Técnicas* – CONICET) were created in 1956, 1957 and 1958, respectively: these three institutions which, jointly with the National Agency for Atomic Energy (CONEA), constitute the backbone of the scientific and technological Argentinian system, were founded during a military regime.

During Frondizi's administration, from 1958 until his overthrow by a military coup in 1962, the government clearly backed scientific and technological development, in spite of rapid disillusionment amongst many intellectuals who supported him. This disillusionment was due essentially to his Congress' bill authorising private universities to offer legal degrees and to him signing contracts with American oil firms that were completely opposed to his ideas, and which were exposed in an influential book written four years earlier (Frondizi, 1954). In fact, the ideology of President Frondizi can be called 'developmentalism', which meant, among other things, the belief that scientific and technological development was fundamental to turn Argentina, as Frondizi wished, into a capitalist developed country. In that sense, there was not so much

---

[25]   On the one hand, the coup was supported by right-wing scientists such as Houssay; on the other hand, left-wing students ('Reformists') had suffered Perón's persecutions and had – after the coup – much influence in the universities. The military government rewarded both groups.

[26]   The poor political relationship between Houssay and the reformist students may be observed, for instance, in the debate in the Council of the Faculty of Medicine of the University of Buenos Aires between Houssay who, as member of the Council had proposed restrictions on the number of students who could be admitted to the Faculty, and the members representing the students, who were opposed to the restriction (Cibotti, 1996). And Houssay's distrust in the 'politicization' of the University is clearly seen in his first speech as a member of the Argentinian Academy of Arts in 1939. In this speech he says of the late academician Ángel Gallardo '[h]e was one of the few men who ruled it [the University of Buenos Aires] who has not been contaminated by the so-called university politics, which often is a fight of egoisms trying to dominate' (Houssay, 1939).

difference between Frondizi and the progressive scientists and university people (the 'Reformists') who had a strong influence in several public universities, especially in the University of Buenos Aires. Here, they controlled its Faculty of Exact and Natural Sciences, which was transformed into an internationally recognised research centre. In any event, Frondizi needed to hold onto power under difficult circumstances. On the one hand, he confronted the Peronists, because he did not achieve what he had promised them in a secret agreement that permitted the Peronists to vote for him, thus guaranteeing his triumph in the 1958 elections. He also had to face right-wing politicians and the Armed Forces, because that same secret agreement was considered a betrayal of the anti-Peronist spirit. Frondizi's tactic was to try to obtain more and more support from the Armed Forces and the Catholic Church. The Church was not particularly interested in scientific and technological progress; its ambition was to control education, to create Catholic universities and to avoid the implementation of projects based on 'subversive' ideas, such as divorce.[27]

As on many occasions in Argentinian history, the military who overthrew Frondizi were, in some sense, a deeply contradictory group. Frondizi had guaranteed support for many scientific and technological advances, including science and technology with military purposes. When Frondizi was overthrown, the Armed Forces controlled the government despite the existence of a civilian President (José María Guido) and they did not yet dare to overthrow the authorities of the public universities. That said, antiscientific measures were taken, such as the dismissal, in the prestigious National Institute of Microbiology, of its director Ignacio Pirosky and of some of his collaborators. Pirosky was a distinguished scientist and his dismissal also revealed a certain anti-Semitic bias. It is equally worth mentioning that, due to the repressive atmosphere created at the Institute, the future Nobel Prize winner César Milstein resigned from his position there and returned to Great Britain, where he continued his brilliant career until his death (Hurtado, 2010).

---

[27]   Although many non-confessional private universities currently exist in Argentina, in the beginning the main beneficiary of Frondizi's policy regarding private universities was the Catholic Church, which was, besides, the institution most interested in this policy. The Church mobilised many people in support of private universities (Ghio, 2008).

The Definitive Break Up between Sectors of the Scientific Community and the Armed Forces

The new military regime, with its civilian President, lasted until 1963, when Arturo Illia was elected President (with Peronism outlawed) and relations between the universities, the scientists and the government improved, in spite of the fact that student demonstrations demanding an increase in the educational budget contributed to weakening the government. In 1966 a new coup overthrew Illia and the military appointed General Juan Carlos Onganía as President. That coup, against a President who always respected freedoms and the rule of law, during whose government the economy grew annually by seven per cent, and who was gradually legalising the Peronist Party, shows a grave and deep distrust of democracy in Argentinian society at that time. That distrust was useful to Catholic fundamentalism and to the groups that would nowadays be called neoliberal, to impose their ideologies over several years. In fact, Illia was overthrown almost without opposition. One of the few opposing voices came from the University of Buenos Aires where its Rector issued a strong condemnation of the coup.

One month later, on 29 June 1966, what was symbolically the most important event of the relationship between the Armed Forces and science and technology in Argentina in the twentieth century took place. This was the so-called Night of the Long Sticks, when the Federal Police, under the orders of an Army general, entered the Faculty of Exact and Natural Sciences at the University of Buenos Aires, beat students, graduates and professors gathered there and detained them for several hours. Although this aggression was incomparably less grave than the repression unleashed by the military in 1976, which also concerned university people and scientists (there were no dead in 1966, only some who were bruised), it has been engraved in the collective memory as a potent symbol of military brutality and of the complete incomprehension and mistrust of the military *vis-à-vis* what science means.[28] As a consequence, around 1,300 professors and teaching assistants of the University of Buenos Aires resigned. Some of them abandoned their scientific activities, others emigrated abroad (mainly to Chile and Venezuela) and others took refuge in the National Council of Scientific and

---

[28]   On the same day the Federal Police burst into the Faculty of Architecture and Urbanism at the University of Buenos Aires and, besides using violence against professors, graduates and students, destroyed mock-ups of buildings prepared by the students. Very significantly, this event shows the feeling of the new military government towards intellectual activities.

Technical Research (CONICET), where they were tolerated by the authorities because they had no direct contact with students.[29]

CONICET was not a refuge for everybody. Despite the fact that Houssay continued being its President, due to his personal and scientific prestige as well as his right-wing ideology, CONICET ruled that any researchers who wished to be admitted to the institution would need to be authorised by the Secretariat of Intelligence of the State (*Secretaría de Inteligencia de Estado* – SIDE). This was an efficient method to exclude people with 'suspicious' ideologies.[30] In fact, in 1970, under Houssay's presidency and one year before his death, the Board of CONICET proposed to the military authorities of the country that they appoint Carlos Alberto Sacheri as Scientific Secretary. Dr Sacheri was the president of the group The Argentinian Catholic City, which published an anti-Semitic, anti-liberal, anti-communist and anti-French Revolution journal.[31] Houssay voted against the appointment because the candidate had no scientific background, but he considered that nonetheless Sacheri had 'a good record and favorable conditions'.

In the non-university scientific and technological institutions (INTI, INTA, CONEA and others) the ideological filter of SIDE also existed, but was seldom used during Onganía's administration. In the military mindset the communist devil materialised in the public universities. It is interesting to observe that there were sectors of the Armed Forces, who clearly recognised the importance of science and technology, who were extremely worried because of the crisis originating in the Night of the Long Sticks (they were also worried when Frondizi was overthrown) and who considered that the exodus of scientists endangered national defence. They were evidently a small minority, and in spite of their pressure and of the enormous public impact of the Night of the Long Sticks (heightened because among those beaten was a distinguished American mathematics professor, Warren Ambrose, who sent a letter to the *New York*

---

[29]   The poor relations between many groups of students and the Armed Forces have a long tradition in Argentina, even before the 1930 *coup d'état*. Halperin Donghi (1962) describes an 'extraordinary fuss' in a lecture at the Faculty of Law of the University of Buenos Aires in 1927, where subjects related to national defence were discussed. More than 150 officers had attended the lecture, as well as the Rector of the University and the Minister of War (the future President Justo). And the 'troublemakers' were students of law, not future scientists. That is, the military mistrusted students before they mistrusted scientists.

[30]   The corresponding resolution was passed by the Board of Directors of the CONICET at the end of 1967 (Hurtado, 2010).

[31]   Editorial. *Ciencia Nueva*, 5, 1970, p. 4.

Times),[32] Onganía did nothing to prevent the exodus of scientists. Although they could not influence the government, those sectors tried to save what could be saved: scientist Marcelino Cereijido (Cereijido, 1990) tells how after the 1966 assault on the University (after which Cereijido was dismissed from his position by the new Rector), he was called by Brigadier Bosch, President of the Board of Scientific and Experimental Research of the Armed Forces. Bosch was not only extremely worried about the potential loss of scientists due to resignations and dismissals, but also offered Cereijido a position. Bosch fully understood Cereijido's explanations of his research on biological membranes and made very precise and accurate comments on the importance of scientific development in the country and, above all, on its concrete applications. Next, Bosch accompanied Cereijido to see the Director of the Institute of Scientific and Technical Research of the Armed Forces, Rear Admiral Milia, who also had very firm ideas about the applications of science for development and national security, and secured a position for Cereijido. According to Cereijido, around one hundred scientists were protected in similar ways by Bosch and Milia.

Analogously, when in 1974 President María Estela Martínez de Perón (Perón's widow) dismissed the authorities of the University of Buenos Aires and replaced them with an extraordinarily reactionary and obscurantist group, I was a witness to how Commodore Vélez, in charge of an area of research at the National Institute of Water Science and Technology (currently the National Institute of Water), protected and offered positions to around thirty scientists who had been fired from the University. Unfortunately, the general political climate was extremely disagreeable and many scientists, in any event, eventually abandoned science or Argentina.[33]

The only successful and relevant military effort at tolerance, and even support, for the scientific and technological community in Argentina happened at the National Agency for Atomic Energy (CONEA), where, as an almost isolated phenomenon in modern Argentinian history, an environment that was practically constant from the last years of Perón's administration onwards

---

[32]   Professor Ambrose's letter, dated 31 August 1966, was published on 11 September. The international impact of the assault against the universities is indicated by the articles that appeared in the *New York Times* in the days following the Night of the Long Sticks.

[33]   This protection of scientists against ideological discrimination lasted until 1976. The dictatorship installed that year mercilessly enforced its discriminatory measures; several scientists and technologists working in public institutions (not only universities) were kidnapped and murdered (or disappeared) and many others were dismissed.

could be observed.[34] Anyway, during the first years of its existence, there was strong opposition by many physicists to the considerable amount of money put aside for CONEA. Those physicists thought that it would be much more useful to use this money for laboratories in national universities. In this regard, after Perón's fall, the prestigious physicist Enrique Gaviola proposed that the CONEA be closed and its buildings, equipment and personnel be transferred to the universities (Hurtado, 2010); Mario Bunge (2009) commented that he was one of the opposing physicists, which, in his case, included a great mistrust *vis-à-vis* the Navy, under which, tacitly, CONEA operated (technically it was not the case, but in practice almost all its Presidents, until 1983, were Navy officers).[35]

After a turbulent period, in which an urban guerrilla movement emerged, the military government, with General Lanusse as President, had to call a general election in which no political party was outlawed, and in 1973 an extremely unsettled time began, in which the Peronist right prevailed over the revolutionary Peronist sector in 1974.[36] As already mentioned, the leftist authorities of the universities were replaced by rightist ones; many professors and teaching assistants were dismissed.[37] Moreover, a Dean of the Faculty of Philosophy, the priest Raúl Sánchez Abelenda, exorcised his Faculty to chase away evil spirits. With these extremely reactionary attitudes, the military had nothing to do: the government had been elected in transparent elections and the protagonists were all civilians, under the extremely strong influence of the Catholic Church. In Argentina, the Church has always played a main role in the attack against free thinking and science.[38]

---

[34]   Probably the most successful fruit of the military support for CONEA is the construction of the particle accelerator TANDAR between 1975 and 1986, described by Hurtado in his book mentioned previously and in Hurtado and Vara (2007). That was a 'big science' project which, although begun and finished during constitutional governments, was carried on essentially during the 1976–83 military dictatorship, which backed the project both politically and economically.

[35]   Personal correspondence with Mario Bunge, December 2009. He comments on this also in the foreword to Bernaola (2001).

[36]   The turbulent period 1973–74 of 'revolutionary' university did not seriously address the scientific and technological problems of the country, probably because, on the one hand, it did not have enough time and, on the other hand, it had to face the political fight against the right; further analysis is beyond the scope of this work.

[37]   In fact, when the military overthrew President María Estela Martínez de Perón in 1976, few professors in the University of Buenos Aires were fired, because the 'dirty work' had been completed beforehand.

[38]   When the University Reform began in Córdoba in 1918, the rallying cry was 'No to the priests!' (see, for instance, Bruera, 2009).

The Dictatorship 1976–1983

In 1976 Perón's widow was overthrown; the last military coup in Argentina started the bloodiest dictatorship of the twentieth century in Argentina. Curiously, many of the scientists who had to flee from Argentina went to Brazil, where the government was also a military dictatorship, but where they could comfortably continue their scientific and academic careers. It is worth noting how much the ways in which the Argentinian Armed Forces tried to impose their ideology had changed by this time: during the 1943–46 military government they changed the lyrics of tangos, when written in slang, so that they become more 'respectable'; during the 1966–73 military government they detained unmarried couples caught in hotel rooms rented by the hour because they were regarded as immoral; during the 1976–83 military government they unleashed indiscriminate repression and thousands of people disappeared. The ideological influence of fundamentalist Catholic fascism in the Armed Forces is discussed, for instance, in Finchelstein (2010). Here, it is interesting to flag the special role played by CONEA. On the one hand, during this dictatorship (as well as during the others) some scientific institutions, mainly CONEA, were strongly supported and their scientists and technologists could work without too much political interference; on the other hand, for instance in CONEA, some scientists 'disappeared' and its President, Admiral Castro Madero, did nothing to save them.[39] CONEA and related institutions, on the one hand, enjoyed a stability and the existence of an environment that are striking in a context which, generally speaking, was so harmful to scientific and technological development. On the other hand, modern mathematics was considered 'subversive'.[40] It is possible that this protection of scientists and technologists working in CONEA is related to secret militaristic wishes: that to have an important war industry, and therefore powerful Armed Forces, modern science and technology are necessary. The cost of having modern 'ecological niches' in some areas, completely isolated from the general state of development of the country, tends to be high. The archetypical example is North Korea, which has the atomic bomb and simultaneously suffers frightening famines. We think that a similar situation in Argentina would be politically unfeasible.

---

[39]   On the website http://ate-cnea.blogspot.com/2011/04/35-anos-del-golpe.html a list exists of the 14 people who disappeared in CONEA.

[40]   Terán (1979) mentions that on 13 December 1978, the Buenos Aires newspaper *La Opinión* reported that the Ministry of Education had formally consulted the National Academy of Exact, Physical and Natural Sciences about the potentially subversive power of modern mathematics.

It is worth mentioning that the contradictions of the Armed Forces were not related exclusively to the scientific and technological communities. In general, the military governments in Argentina, with the exception of that of 1943–46, established economic policies which, in current parlance, we could call 'neoliberal'. But nevertheless, their nationalism led them to prevent any privatisation of public enterprises. In fact, the most neoliberal of the military governments, the 1976–83 dictatorship, not only did not privatise any enterprise, but *nationalised* a very important one: the Italian–Argentine Electric Company.

The Scientific and Technological Community in Democracy

From the democratic restoration in 1983 onwards, the Armed Forces lost their influence over the control of the scientific and technological institutes, and from the 1990s on, they lost all influence in the country. However, only in the last few years, from 2003 on, does there seem to be a state policy that supports science. During Alfonsín's administration (1983–89) and in the second part of Menem's administration (1989–99) there were competent people in charge of the then Secretariat of State for Science and Technology: Manuel Sadosky, a distinguished intellectual and founding father of computer science studies in Argentina, and Juan Carlos del Bello, respectively. Nevertheless, although all ideological discrimination disappeared for people who wanted to obtain a position in the national system of science and technology, the budget for science and technology continued to be very meagre, scientists and technologists continued to earn low salaries, scientific laboratories continued to be poorly equipped, the mean age of researchers continued to increase and a significant proportion of young (and not so young) researchers emigrated, for economic, not political, reasons.

Furthermore, the ambitious project promoted by Sadosky of making up for lost time in the area of computer science, through intensive training of human resources, was interrupted when Carlos Menem became President in 1989, and a large part of the effort was wasted.[41] During the first years of Menem's legitimate and democratic government, the same obscurantist sectors that had occupied

---

[41]   The project consisted in the creation of the Latin American Higher School in Informatics (ESLAI), a high-level university institution which did not survive the change of government. The history of ESLAI may be consulted in Jacovkis (2004) or Aguirre (2003), for instance.

positions of power in all military dictatorships recovered positions.[42] This shows that obscurantism was deeply ingrained in civil society, and in particular in the intellectual sectors of the Peronist right. Those sectors had no influence during Alfonsín's administration, save for the fact that they conserved institutional power and limited the government's room for manoeuvre. Criticism of Alfonsín is more related to his government's indifference to science and technology than to an obscurantist policy, which by no means existed while he was the President. In some respects, given that President Menem was legitimately and democratically elected in 1989 (and re-elected in 1995), his attitude regarding science and technology, especially during his first years of government, is an alarming sign of lack of interest and incomprehension of an important part of Argentinian society regarding the importance of science and technology in a developing country.[43] And, of that phenomenon, one cannot accuse the Armed Forces.

Summing up, one can say that, in Argentina, usually the obscurantist sectors *vis-à-vis* science were also obscurantist regarding education; that the Church was the intellectual force behind them and they obtained more and more power from 1930 on. Also, that they had so much influence in the military that the Armed Forces became the armed wing of national obscurantism. But the Armed Forces had no 'monopoly' on obscurantism. Not all was darkness during military governments and not all was light during democratic governments.

List of References

Aguirre, J., 2003. 'La ESLAI: advenimiento, muerte prematura y proyección'. *Newsletter Electrónica de SADIO*, 8.

Artopoulos, A., 2007. *Proyecto Pulqui II. Una sociología histórica de la innovación tecnológica en tiempos de Perón*. MS Dissertation, Universidad de Buenos Aires.

Babini, J., 1986. *Historia de la ciencia en Argentina*. Buenos Aires: Ediciones Solar.

---

[47]   For instance, Bernabé Quartino, Rector of the University of Buenos Aires between 1971 and 1973, was the President of CONICET at the beginning of Menem's administration (1990–91).

[43]   As a symbolic example, which the scientific community remembers very well, the powerful Minister of Economy during several years of Menem's administration, Domingo Cavallo, contemptuously referred to a female scientist suggesting that she should go and 'wash the dishes'.

Bernaola, O.A., 2001. *Enrique Gaviola y el Observatorio Astronómico de Córdoba.* Buenos Aires: Ediciones Saber y Tiempo.

Bruera, R.L., 2009. *La Reforma Universitaria y el surgimiento de una nueva generación intelectual argentina con proyección latinoamericana.* PhD Dissertation, Universidad Nacional de Rosario.

Buchbinder, P., 2005. *Historia de las universidades argentinas.* Buenos Aires: Editorial Sudamericana.

Cereijido, M., 1990. *La nuca de Houssay.* Buenos Aires: Fondo de Cultura Económica.

Cibotti, E., 1996. 'Bernardo Houssay y la defensa de la Universidad científica en Argentina'. *Estudios Interdisciplinarios de América Latina y el Caribe,* 7(1), pp. 41–56.

Factorovich, P. and Jacovkis, P.M., 2009. 'La elección de la primera computadora universitaria en Argentina'. In: J. Aguirre and R. Carnota (eds). *Historia de la informática en Latinoamérica y el Caribe. Investigaciones y testimonios.* Río Cuarto: Universidad Nacional de Río Cuarto, pp. 83–97.

Finchelstein, F., 2010. *Transatlantic fascism. Ideology, violence and the sacred in Argentina and Italy, 1919–1945.* Durham, NC: Duke University Press.

Frondizi, A., 1954. *Petróleo y política.* Buenos Aires: Raigal.

Galison, P., 2005. *Einstein's clocks, Poincaré's maps.* New York: W.W. Norton.

Ghio, J.M., 2008. *La Iglesia Católica en la política argentina.* Buenos Aires: Prometeo.

Goñi, U., 1998. *Perón y los alemanes.* Buenos Aires: Sudamericana.

Halperin Donghi, T., 1962. *Historia de la Universidad de Buenos Aires.* Buenos Aires: Editorial Universitaria de Buenos Aires. Reprinted in 2002.

Hernández Andreu, J., 1987. 'Una reinterpretación de las crisis económicas mundiales de 1929 y de 1973. Un análisis del sector triguero'. *Revista de Historia Económica – Journal of Iberian and Latin American Economic History (RHE–JILAEH),* 5(1), pp. 99–117.

Houssay, B.A., 1939. 'Discurso pronunciado al incorporarse a la Academia Argentina de Letras'. *Boletín de la Academia Argentina de Letras,* 8, pp. 317–43.

Hurtado, D., 2010. *La ciencia argentina. Un proyecto inconcluso: 1930–2000.* Buenos Aires: Edhasa.

Hurtado de Mendoza, D. and Vara, A.M., 2007. 'Winding roads to big science: experimental physics in Argentina and Brazil'. *Science Technology Society,* 12(1), pp. 27–48.

Jacovkis, P.M., 2004. 'Reflexiones sobre la historia de la computación en Argentina'. *Saber y Tiempo,* 5(17), pp. 127–46.

Jacovkis, P.M., 2006. 'The first decade of computer science in Argentina'. In: J. Impagliazzo (ed.). *History of computing and education 2 (HCE2)*. IFIP International Federation for Information Processing, Volume 215, Boston: Springer, pp. 181–91.

Klich, I., 1999. 'Argentina'. In: *American Jewish Year Book*, Volume 99, New York: American Jewish Committee, pp. 263–75.

Lugones, L., 1979. 'El discurso de Ayacucho'. In: *El payador y antología de poesía y prosa*, Caracas: Editorial Ayacucho, p. 305.

Mariscotti, M., 1985. *El secreto atómico de Huemul*. Buenos Aires: Editorial Sudamericana-Planeta. Fourth printing, Estudio Sigma, 2004.

Ortiz, E.L., 1992. 'Army and science in Argentina'. In: P. Forman and J.M. Sánchez Ron (eds). *National Military Establishments and the Advancement of Science and Technology*. Dordrecht: Kluwer, pp. 153–84.

Potash, R.A., 1981. *The Army and politics in Argentina 1945–1962. Perón to Frondizi*. Stanford: Stanford University Press.

Sanguinetti, H., 2006. *Breve historia del Colegio Nacional de Buenos Aires*. Buenos Aires: Juvenilia Ediciones.

Terán, O., 1979. 'La Junta Militar y la cultura. El discurso del orden'. *Cuadernos de Marcha*, 2, July-August 1979, pp. 49–54. Reprinted in *Cuadernos del Pensamiento Latinoamericano*, 20 (2013). Available at: http://www. cuadernoscepla.cl/?page_id=362.

Vaccarezza, R., 1981. 'La elección del doctor Houssay como profesor titular de Fisiología en la Facultad de Ciencias Médicas'. In: V. Foglia and V. Deulofeu (eds). *Bernardo A. Houssay, Su vida y su obra, 1887–1971*. Buenos Aires: Academia Nacional de Ciencias Exactas, Físicas y Naturales, pp. 177–81.

Zanatta, L., 1996. *Del estado liberal a la nación católica*. Bernal: Universidad Nacional de Quilmes.

Chapter 10

Science Policy in Argentina During the 'Dirty War'

Diana Maffía

In this article I analyse the transition between the short democratic period and the last *coup d'état* in Argentina (1973–76). I am closely involved in this story, since I was an undergraduate student during that time. That passage, from the collapse of dictatorship and the democratic restoration in 1983, convinced me that many evils inhabited society long before the legal breakdown.

'Dirty War' is a euphemism for the last Argentine dictatorship. This expression refers to a period characterised not just by a fracture of the institutional order (as in every dictatorship), but, most specially, by the flagrant violation of human rights. This meant that political objectives were accomplished by means of mechanisms such as torture, disappearances, kidnappings, baby abductions, assassinations and throwing live prisoners into the sea from planes. Compared to the latter, crimes such as shutting universities, proscriptions, censorship or burning of books that constitute the most immediate focus of this chapter seem less horrid.

Talking about the 'Dirty War' implies the existence of two factions; in Argentina this is known as the 'Theory of the two Demons'.[1] According to this theory, there were two sides in the war: the Armed Forces and the guerrilla groups. The large majority of people, marginal to the rationale and actions of each side, became the victims of the 'Dirty War'. Nevertheless, what took place in Argentina was not a war but a State crime, a crime that democracy eventually

---

[1]   In the foreword of the report of the National Commission on the Disappearance of Persons (CONADEP) written by Ernesto Sábato – a report also known as *Never Again*, – he says : 'During the seventies Argentina was convulsed by terror coming both from the extreme right and extreme left [...] to the crimes perpetrated by terrorism, the Armed Forces responded with a terror infinitely worse than the one they were fighting, because from 24 March 1976 they counted on the power and impunity of absolute State support for kidnapping, torturing, murdering thousands of human beings'.

brought to trial in an unprecedented episode when a constitutional government elected by the people judged the crimes perpetrated by the previous dictatorship.

Carlos Nino described this experience in detail, and the philosophy that inspired it, in his book *Juicio al Mal Absoluto* (*Radical Evil on Trial*). There he writes: 'There were few instances in world history, certainly none in Latin America, of prosecuting those responsible for massive human rights violations. Argentina did just this and, even more remarkably, undertook this endeavor without an invading army or a division of the armed forces backing the trials, relying on nothing more than moral appeal' (Nino, 1996, p. 186). The concept of absolute evil or radical evil (taken from Hannah Arendt, who in turn owes it to Immanuel Kant) is of use in this case to refer to a system of total domination, whose abuses cannot be judged from inside, since crime and murder are the rule (Arendt, 1996).

Put somewhat humorously, in the twentieth century Argentina was prodigal with dictatorships. In fact there were six *coups d'état*, in 1930, 1943, 1955, 1962, 1966 and 1976. The last was significant due, sadly, to the high number of people who went missing. It ended on 10 December 1983, which was the Human Rights Day. The newly elected President Raúl Alfonsín chose that day to assume his office. Every democratic government elected in the period from 1930 – when the first *coup d'état* took place – to the end of the last one was overthrown. Twenty-five years of the fifty-three year period involved military governments interrupting intellectual progress, scientific productivity and, in some instances, regional leadership. None of the governments (with different political allegiances) was able to complete its period of office. This is why we value the democratic continuity we have enjoyed from 1983 so highly.

With democracy the intellectual spirit of University Reform took place in 1918. University Reform was an early twentieth-century phenomenon that would leave a strong mark on Argentina's academic life. It produced a critique of the feudal style of the chair system and promoted participative forms of university government. Although these new winds of change were interrupted by the dictatorship in 1930, the reformist spirit allowed, for instance, an extremely young Mario Bunge, aged only 18, to found the Argentine Worker's University (*Universidad Obrera Argentina*) in 1936. The Argentine Worker's University had both professors and workers in its academic council and it offered an education on technical, humanistic, political and union matters. Eventually, Bunge would give up reformism because of its few ties to science. Principles such as critical thought, the foundation of knowledge and the culture of logics were also present in Bunge's valuable journal *Minerva* (1944–45), whose continuity, as with the case of the Worker's University, was frustrated not by dictatorship but by Peronism.

The central thesis of this text is that censorship and persecution, some of the totalitarian ideas that impoverished our academic life and some of the laws restricting participation in university life, were not the product of military dictatorships but originated in civil life, within spaces of the democratic system. This is the reason why this chapter will explore the period before the military dictatorship, immediately after Perón's death in July 1974; the period of the gestation of the snake's egg. I shall refer to the first Peronist government (1946–52), with its clear ties to Franco's Regime, to show how such ideas continued some decades later in the incipient democratic spring that lasted from 1973 until 1976. These ideas popularised during the forties returned during the seventies after Perón's death and again during the last dictatorship.

From the First to the Last Peronism

From 1946, in the early days of Perón's first government, Oscar Ivanissevich was first Secretary and then Minister of Education. Far from reformism, he was busy linking Peronist doctrine to transcendental values such as nationalism and religion, inculcating urbane behaviour in school programmes: politeness, respect, neatness, elegance, general attitude within the school, the family and public spaces (Vaccarezza, 2010). Already in that first period, the government's relationship with the Buenos Aires University, which was in the hands of the opposition, was harsh. In 1947, law 13,031 was passed; it eliminated university autonomy and prohibited all political practice within its premises. In 1954, law 13,031 was substituted by law 14,294 which established a formal connection between the executive and the universities.

Dr Oscar Ivanissevich was a militant anti-reformist who ascribed to a fundamental scheme: Home, Family, Fatherland and God (Sigal, 2002). His designation as controller of the University of Buenos Aires exacerbated the dismissal of professors and expulsion of students, together with other repressive measures (Sigal, 2002). Soon designated as Minister of Education, he transformed the political issue into a police problem and, additionally, as Halperin Donghi (1962) has put it, he used his office to execute both personal and political revenge. According to Sigal 'in a few months, the universities lost 70 per cent of their professors and by 1946, a third of their teaching body; 423 professors were dismissed and 823 resigned' (2002, p. 481). Many of them went into exile, others tried to survive by performing intellectual jobs at the margins of the academy.

Neither university opposition nor political opposition to Peronism were homogeneous; it excluded only expressions of the ultra right and Catholic fundamentalism, but included liberals, leftists and radicals, among others. In September 1955 a military coup overthrew Perón's government and the contradictions arising from the diversity of the opposition were revealed. In the struggle for power, tensions were resolved in favour of the liberal wing; these were recalcitrant anti-Peronists who had consecrated General Pedro Aramburu as President of the Nation, two months before the *coup d'état*. From then on, and up to 1966, there was a succession of elected governments interrupted by military coups. Arturo Frondizi was elected in 1958 and overthrown in 1962, Arturo Illia, invested by democratic vote in 1963, was overthrown by the army in June 1966. Historian Luis Alberto Romero calls this period the time of the 'draw' because of the unresolved social tensions between different national political and economic sectors (Romero, 2001). Thus, a sinister antagonism between the Armed Forces, victorious in 1955, and the working class, organised around proscribed Peronism, gained more and more strength and marked our history during the twentieth century. Peronism antagonised both the dictatorship and the Reformist University. Even today, differences between academia and popular sectors still prevent possibilities for approach and dialogue.

In spite of its dictatorial origins and of the proscription of Peronism, the legal order fostered by the Ministry of Education of the two first *de facto* governments helped the University flourish. The ensuing legislation established full autonomy for the University and allowed for the re-appointment of university staff by means of open procedures for the nomination of teaching staff. This was just what the progressive sectors had been demanding, although, as Mignone (1998) rightly pointed out, it silently introduced authorisation for private institutions to be established.

This prosperity was possible, even in difficult university times, because, especially in the sciences, a number of distinguished individuals had attained considerable research and critical achievements. They included Professor Houssay and his group in the Medical Faculty, the Mathematics Department led by Rey Pastor at the Exact Sciences School of the University of Buenos Aires (some of its most distinguished disciples were José Babini, Gregorio Klimovsky and Oscar Varsavsky) and the Mathematics Research Department under the direction of Mischa Cotlar. Within this environment a large segment of the reformist political activists were scientifically educated. Their scientific quality allowed them to conceive their work overall as an outstanding project which, at the same time, legitimated their authority and action (Oteiza, 1992).

This sort of 'reformist restoration' (Tagashira, 2004) – which occurred after the purge of Peronist faculty – produced, mainly at the University of Buenos Aires, a scientific renewal that lasted 10 years. And, mainly in basic and social sciences, it produced work of international research standard. Among those scientists, it is worth highlighting the modernising role of Rector Risieri Frondizi's work. Frondizi had a clear vision about the interrelations between university and society, and particularly about scientific research. Frondizi assumed his office in the last days of 1957. Among his achievements we can draw attention to the creation of the prestigious University of Buenos Aires publishing house, EUDEBA, raising funds to pay for full time research professors, initiation of the construction of the campus of the *Ciudad Universitaria*, the creation of faculties, departments, laboratories and seminar rooms, among other contributions that improved Argentine science and scientific education (Frondizi, 1971). Many of these improvements were achieved by the University in total isolation *vis-à-vis* the tremendous economic and institutional problems the country was undergoing, a situation that generated internal criticism within the University around the role of science and intellectuals in society. It also produced memorable debates, for instance, the discussion that gave rise to the dispute between scientism and antiscientism.[2] The scientism-antiscientism dispute encapsulates the confrontation between a universalist and neutral conception of science and one which promotes a politically involved, national and regional orientation.

Undoubtedly, Oscar Varsavsky (1994), who had followed the reformist movement at the Exact Sciences Faculty from its beginnings, was a passionate representative of the scientism-antiscientism debate. He advocated a university free from 'fossils', by which he meant professors incapable of doing research. Still, he warned against scientism, a more powerful danger than academic lack of aptitude. Scientism was the attitude assumed by the researcher who forgets his social responsibilities in order to progress in his or her scientific career. Varsavsky addressed the geopolitical division in scientific research with remarkable vision: the relationship between centre and periphery; the privileges enjoyed by powerful rich countries; the scientists who enjoyed grants in those countries but, on returning home, did not engage with local social issues for fear of losing prestige. The neutral, meritocratic appearance of contests based on

---

[2]   A fundamental example of that debate was published in 1975 in the Collection *Ciencia e Ideología. Aportes Polémicos*. Buenos Aires: Ed. Ciencia Nueva, that publishes works by Gregorio Klimovsky, Oscar Varsavsky, Jorge Schvarzer, Manuel Sadosky, Conrado Eggers Lan, Thomas Moro Simpson and Rolando García.

academic background or quantitative evaluation was unmasked by Varsavksy as conservative resistance and protection of hegemonic interests.

Through his critique, Varsavsky generated strong opposition within both conservative and progressive sectors. The latter sector shared his vision about the need to attain independent economic cultural and political development but, differently from him, had faith in the adequacy of scientific forms of evaluation. Thus, they argued that abandoning those criteria would result in a lack of objective criteria, leading to arbitrary, irrational and fanatical decisions. On the other side, Varsavsky deplored the under-politicisation that characterised his colleagues, as they opted to become the ostensibly neutral scientific base for national development (Varsavsky, 1994).

The debate came to an end brutally in 1966 with the euphemistically called 'Argentine Revolution'; the Regime imposed the National Security Doctrine under the *de facto* governments of Juan Carlos Onganía (1966–70), Roberto M. Levingston (1970–71) and Alejandro A. Lanusse (1971–73). To this dictatorship we owe the Night of the Long Sticks [*Noche de los Bastones Largos*]. On 29 July 1966, Buenos Aires' public university suffered the most virulent ideological attack from the State; infantry troops entered the university cloisters and began physical repression of teachers and students. That night the dismantling of higher education began. Political exile for full groups of teachers, research teams were disbanded, ideological polarisation was induced from above in order to void the academic cloisters of politics. All these actions destroyed, in months, ten years of academic and scientific–technological growth. Ninety per cent of the professors in the Exact Sciences Faculty resigned, more than a thousand faculty members fled the country or remained outside public education.

The Institutional Order Previous to the *Coup d'État*

The restoration of the institutional order in 1973 brought with it important news: the end of Peronism's proscription and, eventually, the party's electoral victory. Maybe because of Peronism's double polarity and antagonism *vis-à-vis* the dictatorship, on the one side, and liberal ideas on the other, the party set in motion an educational project that, after Perón's death in 1974 and during the constitutional period (1973–76), became extremely contradictory. With respect to academic policy it is necessary to distinguish two very different – almost antagonistic – management styles: that of Minister Jorge Taiana, who was in office up to August 1974, and that of Oscar Ivanissevich up to the *coup d'état* on 24 March 1976.

When the popular government came into power and during its first period, the University of Buenos Aires became, in contrast, the National and Popular University of Buenos Aires – with the intention to prevent liberal reform. The connection between professors, students and workers was promoted; the University was placed at the service of the Peronist economic, political and cultural plan. Syllabuses were modified by the Chairs, and 'National Chairs' were created. In consonance with its popular character, the access requirements for the University were lowered, thus generating massive attendance that was also intended to be more inclusive. Community programmes related to marginal sectors were stimulated, students and teachers participated in popular alphabetisation and political education programmes in working class neighbourhoods.

But things are not always what they appear to be. Minister Taiana was in power for only 15 months and, although his name is associated with a progressive university open to the people through unrestricted admission and popular chairs, the truth is that this was a 'normalization'; that is, his exercise of authority was not in accordance with the University's regulations. There was no Rector but instead a Controller and the University Assembly and the Rector's Council were respectively replaced by the President of the Nation and the Ministry of Education.

The University Law was passed on 26 March 1974. In its 5th article this law forbids political activity. The text reads: 'The proselytism of political parties or of ideas contrary to the democratic system characteristic of our national organization remain prohibited in the domain of the university'. The same law applied to the professors, who risked their position if they did not comply (art. 11°). Another rule that had extremely bad consequences allowed the executive to intervene in the universities in case of a breach of the public order, or of subversion against State power (art. 51°). Among the progressive aspects, was the guaranteed participation of students and non-teaching university staff in university politics (arts 27° and 33°), and the reincorporation of the professors dismissed for political reasons between September 1955 and May 1973.

When Perón died on 1 July 1974, it was obvious that Taiana's days were numbered. The students of the University of Buenos Aires, faced with the prospect of a political backlash that eventually took place, occupied the Dean's office and eleven schools. Their achievements and the opening up of university life, which was defined at the time by the government as 'disorder and chaos to be overcome', brought about Taiana's dismissal from the Ministry of Education and his replacement by Oscar Ivanissevich, and as the Rector of the University of Buenos Aires, the ominous figure of Alberto Ottalagano was appointed. That was when night fell: state terrorism and the Triple A were the preamble for the dictatorship. The 'Ivanissevich Mission' and Ottalagano's intervention into the University of Buenos Aires initiated the 'cleaning' of problematic teachers, who were dismissed, and the harsh repression of the student movement. Ottalagano imposed undemocratic measures by using democratic laws, even Taiana's University Law. According to Ivanissevich, Taiana's University Law was the instrument that provided the legal framework regulating the functions of University controllers.

As the breach of public order allowed the executive to intervene in educational institutions, Decree No. 865 formalised Ottalagano's intervention in the universities, which was then followed by their closure. One of the Ivanissevich mission's main tasks was ideological purging. For that purpose, he used another already existent mechanism: the 1973 Dispensability Law that suspended every provisional university staff appointment (teachers and non-teachers) before the closure of the University; that is, any contracts were rendered ineffective. The spate of sackings would eventually reach staff members who had passed competitions, but whose nomination had not yet been confirmed, who were in their posts before October 1974. According to the unions 2,500 teaching posts were lost (De Luca, 2009).

Ottalagano was proud of his fascist sympathies. He believed that his prime duty was to rescue the University from the chaos and anarchy in which the proselytism of communists and foreigners had left it. At the end of this period, he would say that 'the university had been previously the centre for the recruitment of guerrillas and we have brought peace into it'.[3]

When the faculties were reopened in 1976 all kind of police practices had been adopted. To enter, students were forced to show their student cards indicating the schedules for their classes. They were not allowed to stay on campus except when they had classes. Women had to show their handbags. Recording equipment was prohibited because it was forbidden to record classes. Lecture notes had to be taken by hand of the scarce circulating contents and then these scarce items were circulated.

What, then, was the difference between this state of affairs, its discourse, norms and decisions, and the dictatorship? A very important one, democratic order was in force; but ideologically there was none. The setbacks added up. The last faculty to reinitiate its activity was the Philosophy and Literature Faculty,

---

[3]　　Diario *La Opinión*, 19 December 1974.

to which I belonged. There, the purge was difficult on staff because of the great influence of Marxism. This is why authorities proceeded to shut down the most problematic subjects, namely Sociology, Psychology and Education.

Then, in what constituted an important reversal in terms of the University's popular orientation, an admission year was reinstated for all subjects, along with a quota system privileging technological disciplines to the detriment of the humanities. Thus, a system of persecution and censorship was first deployed. The horror, destruction and death brought about by the military dictatorship ensued. In 1975 the University was shut. Although exile is usually associated with dictatorship, various academics began to be exiled from before March 24 1976.

Right-wing Peronism's political project had close links to franquism. Juan Fernández Krohn, a Spaniard who brags about belonging to the side that won the Spanish Civil War in 1936, mentions a visit to Argentina to preach to the San Pío X brotherhood during the first period of Videla's dictatorship. Referring to the Spanish fascist Javier Iglesias – connected to the left wing of Peronism and murdered during Menem's presidential period – Fernández Krohn was in favour of the America's poor. From his ideological position, close to Franco and in favour of the dictatorship, he believed that the Peronist in 1975 already announced its allegiance to the American revolutionary cause, through the revolutionary way. This new Peronist stage brought its partition into two deadly opposed factions, the Peronist right and the Peronist left in a crescendo of violence under Isabelita (María Estela) Martínez de Perón, widow of the general. Under her presidency there started what journalists would call 'Dirty War' in the times of the Videla and Viola Military Juntas (Fernández Krohn, 2010). On the other hand, Lefebvre's collaborators in Argentina favoured the military regime unconditionally, they were prominent Peronists (of nationalist extraction) some of them were near to the 'blue' (pro-Peronist) group of the Argentine Army (Fernández Krohn, 2010).

The old Ivanissevich did not want sciences to develop in Argentina. He said that science was expensive and that the researchers of the National Council of Scientific and Technological Research (CONICET) had not invented anything. The dictatorship did not want to support science either; it seemed to be an enemy of the conservative visions they advocated. This controversy between scientism and antiscientism, at the end of the 1960s and beginning of the 1970s, still valued the role of science in a country's development; it only focused on the type of science to be undertaken, the kind of compromises involved, reflection around priorities and consideration of the issue of in whose service and whose interests science should proceed.

The University of the Catacombs: A Shining Experience amidst Dark Years

To conclude, I just want to highlight a shining experience amidst dark years. This was the so-called 'University of the Catacombs'. It was sustained by those scholars and intellectuals who were dismissed or resigned their chairs on the Night of the Long Sticks. They did not resign their vocation and generously kept their critical spirit alive and transmitted it to us, the young who were then unable to find that spirit and those ideas in the impoverished academic cloisters.

I was fortunate to receive my education from the University of the Catacombs, in the Argentine Society for Philosophical Analysis (*Sociedad Argentina de Análisis Filosófico* – SADAF) whose mission was to disseminate analytical philosophy. It was founded by Gregorio Klimovsky, Thomas Moro Simpson, Eugenio Bulygin, Carlos Alchourron, Felix Schuster, Eduardo Rabossi and various other intellectuals who resisted the military dictatorship. There, in a seminar on Human Rights led by Eduardo Rabossi and Carlos Nino, we discussed the Interamerican Commission Report based on its 1979 inspection (*Comisión Interamericana de Derechos Humanos*). There, we considered the ethical implications of the 'Fair War' (*Guerra Justa*) during the Falklands War. Also, the conditions for a trial of dictators and the different degrees of political and moral responsibilities were debated there. A grant system was created at SADAF to compensate those who had had to leave the University, so that they could prepare for the competition in their specialty. Additionally, the Society designed a postgraduate course on analytical philosophy to provide a systematic specialised education so that we would adequately be able to assume the responsibility of reconstructing university programmes after the return of democracy. And, considering other less evident hegemonies, it was there that the first conferences on feminist philosophy were offered by Maria Lugones, and where the first seminar on feminist philosophy was held, which had a strong influence on the development of academic gender studies. It was from there that important officers were recruited during the eighties' difficult democratic transition. SADAF was also the origin of the philosophical foundations of the Trial of the Junta, such an unprecedented event.

I owe so much to SADAF, I owe so much to the conviction of those who were dismissed or resigned their posts on the Night of the Long Sticks. During the years of internal exile they did not resign their vocation and generously kept their critical spirit alive until the restoration of democracy in the eighties. They maintained their passion for knowledge, ethical earnestness as a framework for research, engagement with democracy and human rights. They convinced us – who were then so young – to persevere with our ideas. They had known times in which the collective production of knowledge had been a reality and they thought that the future should find us prepared. As a survivor of a generation that went missing, I assume the power to sustain their conviction and to tell those who are young now that this future is now.

***

To summarise, the intention of this analysis is to throw some light over a sinister period in the history of Argentina: formally democratic but in which all institutions were crumbling, and which at the same time was consolidating the political model which gave ideological support to criminal repression. There are both ruptures and continuities between democracy and dictatorship, and those continuities are frequently shaped by the enforcement of laws that give a seemingly institutional appearance to governments that have come into power illegitimately. Recently, when speaking before a court that was judging him for crimes against humanity, dictator Videla justified his bloody military dictatorship arguing that his actions were legitimated by a decree of the antecedent constitutional government, signed on 6 October 1975 by Italo Luder. This decree enabled the Armed Forces to 'annihilate subversion' and subordinated the police to the army. Similarly, the laws dictated by that government on education and the universities served to deepen dictatorial fundamentalism and to keep us apart from the possibility of an autonomous scientific policy for a period of twenty-five years.

List of References

Arendt, H., 1966. *The Origins of Totalitarianism. New York: Harcourt, Brace & World.*

De Luca, R., 2009. 'El progresismo en tiempos de revolución. La universidad argentina de Taiana a Ivanissevich'. In: *El Aromo* (January–February 2009). Available at: http://www.razonyrevolucion.org/textos/elaromo/secciones/Educacion/deluca46.pdf [Accessed 1 August 2013].

Fernández Krohn, J., 2010. 'El caso Javier Iglesias, o la memoria escindida del peronismo argentino y de los falangistas españoles'. In: Las Crónicas de Juan Fernández Krohn [blog] 30 July 2010. Available at: http://blogs.periodistadigital.com/juanfernandezkrohn.php/2010/07/30/p276730 [Accessed 1 July 2013].

Frondizi, R., 1971. *La Universidad en un mundo de tensiones.* Buenos Aires: Paidos.

Halperin Donghi, T., 1962. *Historia de la Universidad de Buenos Aires.* Buenos Aires: EUDEBA.

Mignone, E., 1998. *Política y Universidad.* Buenos Aires: Lugar Editorial.

Nino, C.S., 1996. *Radical Evil on Trial.* New Haven: Yale University.

Oteiza, E., 1992. *La Política de Investigación Científica y Tecnológica Argentina.* Buenos Aires: Centro Editor de América Latina.

Romero, L.A., 2001. *Breve historia contemporánea argentina.* Buenos Aires: Fondo de Cultura Económica.

Sigal S., 2002. 'Intelectuales y peronismo'. In: Juan C. Torre (comp.), 2002. *Nueva Historia Argentina VIII (Años peronistas 1943–1955).* Buenos Aires: Ed. Sudamericana, pp. 481–522.

Tagashira, R., 2004. 'Oscar Varsavsky y la Facultad de Ciencias Exactas de la Universidad de Buenos Aires entre los años 1955 y 1966'. In: http://rapes.unsl.edu.ar/Congresos_realizados/Congresos/IV%20Encuentro%20-%20Oct-2004/eje6/29.htm [Accessed 1 July 2013].

Vaccarezza, V., 2010. '"Evita me ama": La propaganda política peronista en el sistema educativo'. In: *Suite 101. Net* [blog] 18 June 2010. Available at: http://politicaargentina.suite101.net/article.cfm/evita-me-ama_[Accessed 1 August 2013].

Varsavsky, O., 1994 [1969]. *Ciencia, Política y Cientificismo,* 8th edition. Buenos Aires: Centro Editor de América Latina.

Appendix

History of Science in Spain, Italy and Argentina

As with the scientific research that we have described in this collection, the history of science is not written in a vacuum. In order to situate the different authors in this book within their own national, institutional and disciplinary contexts, it is helpful to characterise the current state of the History of Science as a scholarly activity in each country that we cover. The editors approached all of the contributors to provide some thoughts on the development of their own discipline and this appendix presents a synthesis of these short narratives.

In Spain, History of Science has a long tradition, although an institutionalised and professionalised group of historians working on this topic could not be identified until the second half of the twentieth century. That said, the 'controversy about Spanish science' was one of the main intellectual, and ideological, debates in the last quarter of the nineteenth century. History of Science advanced towards institutionalisation with the creation, in 1934, of the Association of Spanish Science Historians. Members included distinguished historians such as Eduardo García del Real (1870–1947), who held the first chair in the History of Medicine at the University of Madrid, Guillermo Folch (1917–87) a great historian of Pharmacy and Professor at the Complutense University of Madrid and José María Millàs Vallicrosa (1897–1970), who promoted the History of Science from his chair in Hebrew and Arabic at the University of Barcelona, and managed to establish the oldest research school in the discipline in Spain.

After the Civil War, as Professor of the History of Medicine in Madrid, Pedro Laín Entralgo (1908–2001) contributed to the establishment of the History of Medicine as part of the curricula of the Faculties of Medicine (also of Pharmacy), facilitating the creation of academic positions in the area. From 1948 he published the journal *Archivo Iberoamericano de Historia de la Medicina y Antropología Médica* (today, *Asclepio. Revista de Historia de la Medicina y de la Ciencia*). One of Laín's disciples, José María López Piñero (1933–2010), was one of the leading and most influential figures of the professional History of Science in Spain from the 1960s.

After successive attempts from 1973 onwards, the *Spanish Society for the History of Science and Technology* was finally created in 1976 in Madrid, which has published its own journal, *Llull*, since 1977. Around this time, numerous scientists, whose work in the History of Science had formerly suffered contempt from classical Professors of Sciences, were gaining professional recognition as proper historians of science. This culminated in its final recognition in the University Reform of 1983. In 1981 a new journal devoted to the History of Medicine and the Sciences was established, *Dynamis*, with joint editorship from the Universities of Granada, Barcelona, Cantabria and Alicante. In 1991, the Catalan *Society for the History of Science and Technology* was born as an affiliate to the Institute of Catalan Studies. A useful tool is the Bibliography of the History of Science and Technology in Spain (*Bibliografía Histórica de la Ciencia y la Técnica en España*), a cumulative bibliography compiled by the Instituto de Historia de la Ciencia y Documentación López Piñero, at the University of Valencia, registering publications in Spain or by Spanish authors between 1988 and 2009.

Italy as the home country of Galileo has been at the heart of the development of science. But science became an object of historical inquiry around the turn of the twentieth century, thanks to a number of historically – and philosophically – minded scientists, such as mathematicians Federigo Enriques and Vito Volterra. A number of journals dedicated to the history of specific sciences were launched during the first decades of the twentieth century. The *Società Italiana di Storia della Scienza* (http://www.storiadellascienza.net) is today the main learned society in this field. The universities of Pisa, Bologna and Bari offer PhD programmes in the History of Science. The University of Bologna hosts the International Centre for the History of Universities and Science (CIS). Pancaldi notes that as late as 1967 the history of science had very weak institutional foundations, but by 2010 had grown to around 120 tenured positions in the country (Pancaldi, 2010).

As is the case in many countries, we may roughly distinguish two orientations towards the History of Science in Italy. One is what we may designate as the 'internal' history of science, concerned mainly with the theoretical developments of a given field or discipline (usually 'hard' or 'exact' sciences like physics, mathematics, etc.) and done either by scientists of that field interested in its history or by historians specialised in that field. The other seeks to integrate the history of science in wider historical, political and cultural patterns and sits somewhere between the history of science narrowly interpreted and the history of Italian society as a whole.[1] This has been especially true for the more recent period, as the history of science in Italy cannot easily be disentangled from events such as the country's unification or the Fascist dictatorship. In this time frame, the social sciences (for example, economics, demography, sociology) or 'hybrid' disciplines such as statistics or eugenics/genetics have been the object of close attention. An impressive synthesis of this whole line of work has been offered recently by the publication of *Storia d'Italia. Annali 26. Scienze e cultura dell'Italia unita* edited by F. Cassata in 2011.

In Argentina, History of Science, as with many other areas of science, has suffered with the political instabilities of the country. The first Argentinian specialist was José Babini (1897–1984), an engineer turned historian of science and applied mathematician and one of the founders of the *Argentinian Mathematical Union*, the professional society of Argentinian mathematicians. Babini collaborated with Julio Rey Pastor (1888–1962), the Spanish mathematician who, after arriving in Argentina in 1917, practically created the school of mathematics in the country, and who was also interested in the history of science. Anti-Semitism and war in Europe brought two important specialists to the country, Aldo Mieli (1879–1950) and Desiderio (Dezso) Papp (1895–1993). Babini, Mieli and Papp may be considered the founding fathers of History of Science in Argentina.

Babini taught history of science in Santa Fe at the University of Littoral until he was fired after the 1943 right-wing military *coup d'état*; at the same time, the Institute of History and Philosophy of Science that he had created at that university was dissolved. The publication of the journal *Archeion*, which Mieli continued to publish from Argentina after his exile from Europe, was interrupted (it was revived in Europe in 1947 with the name *Archives Internationales d'Histoire des Sciences*). Mieli died in 1950, and Papp migrated to Chile in 1961.

Babini published the first book on history of science in Argentina in 1949, *Historia de la ciencia argentina*, and later wrote many other books in this field, many of them under very difficult circumstances. After another military *coup d'état* overthrew President Perón in 1955 in the context of an extremely polarised society, he was appointed Professor at the School of Sciences of the University of Buenos Aires, began teaching an elective course in History of Science and created a Department of History of Science at the University. He was Dean of this University (1956), member of the National Council of Scientific and Technical

---

[1] While the division is familiar across the academic discipline, its peculiar roots in Italian politics and culture (particularly the history of science as part of a modernising programme in the face of idealism, instrumental views of science and conservative Catholicism) is discussed by Pancaldi (2010).

Research – CONICET – (1958) and First President of the Editorial Board of the University of Buenos Aires, EUDEBA (1958). But in 1966, after another military *coup d'état* which ousted President Illia, the Department of History of Science was also dissolved. Only in 1983, after a democratic government was established in Argentina, did research in History of Science begin to grow again.

The institutionalisation of History of Science as an area of research was slow. Eventually, some institutes or centres in Argentinian universities were created, such as the Centre for Studies of History of Science and Technology 'José Babini', at the University of San Martín, where the journal *Saber y Tiempo*, the only journal of History of Science in Argentina, was published (not without difficulties). Many Argentinian historians of science come from science and medicine, although of course others have a background in philosophy and history. The particular interest of scientists in the discipline is perhaps due to the fact that the political history of the country has so strongly influenced (and often hindered) scientific activities that it was usually very difficult for Argentinian scientists to work in isolation from political conflicts. And so, some of them increasingly became interested in the historical development of their discipline (and of related disciplines) and in the obstacles science encountered.

List of References

Pancaldi, G., 2010. 'Purification Rituals: Reflections on the history of science in Italy'. In: Mazzotti, M. and Pancaldi, G. (eds). *Impure cultures. Interfacing science, technology, and humanities.* Università di Bologna, Dipartimento di Filosofia, Centro Internazionale per la Storia delle Università e della Scienza.

Index

Academic freedom 2, 4, 7–8, 10, 13, 17,
    27–8, 30, 32–3, 41, 84, 94–5, 142
Academy of Italy 151
Achúcarro, Nicolás 67
Adams, John Bertram 133–4
Advisory Committee for the Studies on
    Population (*Comitato di consulenza
    per gli studi sulla popolazione* –
    CCSP) 168
Aguilar, José 131
Air Force 124
Albareda, José María 17–19, 87, 91, 108
Alfonsín, Raúl 182, 194–5, 200
Alfonso XIII 67
Alfonso el Sabio Centre (CSIC) 91
Alonso de Herrera Centre (CSIC) 91
Ambrose, Warren 190
anarchist revolution 70
Anglo-Saxon statistics (or English school)
    xix, 19, 159–60, 163, 170, 172–4
Aniel-Quiroga, José M. 133
anti-national policies 71
Anti-Semitic 16, 188, 190, 213
Anti-Spanish 71, 83, 104–5
applied science 39, 90–1, 142, 144
Argentina x, xiii–xvi, xix, 4–6, 8–10, 12, 14,
    20–1, 59, 71, 73–4, 107, 179–80,
    182–4, 185–95, 199–200, 207–9,
    211, 213–14
Armenteros, Rafael 11, 132–3
army xix, 15, 36, 61, 69, 106, 112, 146, 150,
    152, 155, 182–3, 185, 189, 200,
    202, 207, 209
Association of Liberal Doctors 106
Astronomic Observatory 180
Athenaeum of Seville 86

autarky xvi, xviii, 11, 13–14, 21, 80–1, 86,
    92, 124, 129, 141, 143, 145, 147,
    151–4, 156–7, 171–2, 174
Autonomous University of Madrid
    (*Universidad Autónoma de Madrid*)
    75
autonomy 2–3, 8, 10, 13, 16, 21, 38–9,
    42–5, 51, 94–5, 141, 143, 150,
    160–1, 173, 201–2
Avitaminosis 111–13
Avogadro, Amedeo 7

Badoglio, Petro 152–3, 155
Balanzat, Manuel 73
Banissoni, Ferruccio 155
Barinaga, José 69, 73
Bassols, Narciso 74
Battista, Giovanni 19
Bayesianism 161, 164
Belgrano, Manuel 184
Bernardini, Gilberto 134
Bignone, Reynaldo 183
biomedical research 67
Blanc, Gian Alberto 146, 148
Board for Advanced Studies and Scientific
    Research (*Junta para la Ampliación
    de Estudios e Investigaciones
    Científicas* – JAE) 5, 11, 18, 37–45,
    47–53, 63, 66–70, 73–4, 79–80,
    84–6, 88, 90, 92–3, 103–4, 107,
    114
Board of Aid to the Spanish Republicans
    (*Junta de Auxilio a los Republicanos
    Españoles* – JARE) 73–4
Bolivar, Ignacio 40
Bosch, Carlos 191
Bottai, Giuseppe 150, 152

boundary between science and politics xvii,
　3, 27–9, 41, 43, 51
Brain Physiology Laboratory (*Laboratorio
　de Fisiología Cerebral*) 67
Buenos Aires 73, 179–80, 183, 186–93,
　201–3, 205–6, 213–14
Bureau for Inventions and Research
　Initiatives 142

Cabrera Felipe, Blas 42, 52–3, 60, 67–8,
　71–2, 75
Cabrera Sánchez, Nicolás 75
Cajal Institute 67, 103–9, 114
Cajal school 107–8, 115
Calletti, Pio 153
Cannizzaro, Stanislao 7
Capitalism 145
Cárdenas, Lázaro 74
Carracido, José Rodríguez 31, 35–6, 40
Carrasco Garrorena, Pedro 59, 71–2
Castillejo, José 40, 45–8, 50, 69, 72, 107
Castro, Américo 71
Castro Madero, Carlos 193
Catalá, Joaquín 134
Catalán, Miguel 53, 60, 68–9, 73
Catholicism 10, 12, 21, 30–3, 53, 62,
　64, 81–91, 94–8, 103, 108, 124,
　181–3, 185, 188–90, 192–3, 202,
　213
Cavallo, Domingo 195
Cavanilles, Antonio J. 30
Central Commission for the Examination
　of the Inventions of the CNR 155
Central Institute of Statistics (*Istituto
　Centrale di Statistica* – ISTAT) 163
Centre for Aeronautical Research (*Centro
　de Ensayos de Aeron*áutica) 60
Centre for Historical Studies (*Centro de
　Estudios Históricos* -JAE) 44, 52,
　63, 66–7
Cereijido, Marcelino 191
chemistry 20, 36, 52–3, 63, 67–8, 70, 86,
　90, 108, 127, 142, 145, 156
civil servants 72, 75

Civil War xvii, xviii, 5–7, 10–11, 21, 27, 29,
　39–40, 44, 53, 59, 61–2, 66, 69–71,
　74–5, 79, 80, 83–4, 87, 103–4,
　106–7, 110–15, 122–4, 135, 207,
　211
Cold War xvi, 3, 14, 28, 98
College of Mexico (*Colegio de México*) 74,
Committee for Civil Mobilization 148
Committee for Marine Geography 156
Committee for the Study of Population's
　Problems (*Comitato italiano per lo
　studio dei problem della popolazione*
　– CISP) 167–8
Communist 5, 12, 16, 70, 73, 83, 190, 206
　Communism 83, 183
Conservative government 33, 39, 41–4, 46,
　61, 183
Conservative Spain 69
Convergence with Europe xvii, 59–61,
　64–5, 67–9
Corminas, Ernesto 73
Corporatist model 145
Cossi, Colonel Aurelio 148
Cossío, Manuel 31, 37, 40, 45
Cosmopolitanism 7
Costa, Joaquín 64,
Council of Trent 64
Counter-Reformation 64
Cuba 35–6, 65

Dakin, Samuel A. 133, 136
Dallolio, General Alfredo 148
Darwin, Charles 34, 180
De Castro, Fernando 67, 106, 108
De Castro, Honorato 72
De Gregorio Rocasolano, Antonio 69
De los Ríos, Giner 31, 33–4, 37, 40, 46
De los Ríos, Fernando 63, 68, 72
De Maeztu, María 52
De Maeztu, Ramiro 62
De Morvilliers, Masson 30
De Rose, François 125
De Unamuno, Salomé 135
De Zulueta, Antonio 53, 69, 71
De Zulueta, Luis 47, 60, 71

Del Bello, Juan Carlos 194
Del Campo, Ángel 60, 67, 70
Del Río Hortega, Pío 67, 71, 73
democracy xvi, 1–6, 15, 59, 61, 75, 121,
    189, 194, 199–200, 208–9
    democratic reforms 70
    democratic government 32, 194–5,
        200, 214
Demorazza (Direzione generale della
    demografia e della razza) 168–9
Denina, Carlo 30
Department of Demography and Race
    (*Direzione generale della demografia
    e della razza-Demorazza*) 168–9
Dewoitine, Émile 186,
dictator 17, 75, 209
    dictatorship xiii, xvi, xviii, xix, 1, 3, 5–6,
        8, 12, 15, 19, 43–4, 61, 67–8, 98,
        107, 109, 114–15, 124, 179, 182,
        192–3, 199–202, 204, 206–9, 213
    dictatorship of Primo de Rivera 43–4,
        67–8, 107
Dirty War xix, 199, 207
disaffection 71
Dissolving Ideologies 61–2, 73
domestic production 145
Duperier, Arturo 59, 69, 71, 74

Echegaray, José 31–2, 35, 40
economic autonomy of Italy 143
education 2, 8, 14, 18, 31–5, 38, 41, 51,
        66–9, 71, 75, 82–4, 86–7, 92, 107,
        126, 135, 142, 150, 152, 179–80,
        184, 188, 193, 195, 200–205,
        207–9
    educational missions 68
El Socialista 63
Engine National Institute of Naples 155–6
Enlightenment 64, 83
Escoubés, Bruno 135
eugenics 163, 213
Europe xiii, xiv, xv, xvii, 8–9, 19, 30–1, 38,
    52, 59–62, 64–8, 75, 80, 83, 88,
    129, 149, 213
    European Community 129

European Institutions 71
European Organization for Scientific
    Research 121
European Standards 33, 60, 69
Europeanization 64
Europeanized Spain 64
evolution 34, 164
exile xvi, xvii, xviii, 4, 11–12, 21, 59, 61,
    73–6, 80, 85, 87, 92, 99, 105–6,
    111–12, 114–15, 123, 132, 202,
    204, 207–8, 213
Experimental Centre for the Applications
    of Psychology 155
Experimental Centre of Torre Chiaruccia
    155
Experimental High School (*Instituto-
    Escuela*) 52, 63, 68, 70, 92

faith xviii, 32, 53, 62, 81, 87–8, 154
Falangist 69, 87, 96
Falklands War (Guerra de las Malvinas)
    182, 208
false doctrines 62, 72
Fascism xvi, xviii, 6–8, 10–13, 16, 18–19,
    21, 62, 69, 81–2, 96–7, 104, 124,
    141, 143–5, 157, 159–60, 163,
    168–9, 172, 175, 181, 183, 193,
    206–7, 213
Fernández Krohn, Juan 207
Fernández, Obdulio 67
Fifth Column 60
First World War 19, 28, 141, 148
foreign study grants 36, 38, 42, 46–7, 79
Francisco Franco awards 93
Franco, Francisco xvii, 60–2, 70, 72, 75–6,
    80–1, 85, 86–7, 93, 104, 111,
    114–15, 122, 124, 207
Francoist 69, 124
    Franco regime xvii, xviii, 6–7, 10–11,
        14, 29, 62, 79, 81–4, 87, 99, 103,
        106–10, 114, 122–3, 130, 201
    Francoist Spain 4, 21, 59, 71–2, 106,
        114, 123
    Francoism xviii, 14–15, 49, 53, 75,
        121–4

Ferretti, Pericle 155
fraud
　　electoral fraud 181, 185
　　scientific fraud 2, 4–5, 17
Free Institution for Teaching
　　(InstituciónLibre de Enseñanza –
　　ILE) 33, 38, 44–5, 47, 60–6, 69, 74,
　　84–5, 92, 107
freedom of scientists *see* academic freedom
freedom of thought *see* academic freedom
Frondizi, Arturo 187–8, 190, 202–3

Galvani, Luigi 7, 165–6, 170
Gallardo, Ángel 187
García Alix, Antonio 66
Gaviola, Enrique 192
Gemelli, Agostino 155

Gener, Pompeyo 35
General Assembly of the Spanish Red Cross
　　107
General Council of Medical Colleges 107
General Chemistry Laboratory (JAE) 52
General Physiology Laboratory (JAE) 52
genetics 20, 69, 163–4, 213
Geneva Conference on the Peaceful Uses of
　　Atomic Energy 125
geometry 65
German invasion 156
Giannini, Amedeo 143, 144, 149, 152
Gimeno, Amalio 38, 40, 43, 48–9
Giner, Hermengildo 34
Giner de los Ríos, Francisco 31, 33–34,
　　37–8, 40, 46–7, 63
Giral Pereira, José 72
Giral González, Francisco 59, 61, 75
González, Joaquín B. 180
Gould, Benjamin 180
government of science 16, 37, 44
Gregory, Bernard 135
Guido, José María 188
Gutiérrez, Juan María 180

Hampton, George 131
heliocentrism 65

Hernández Tomás, Jesús 70
Hernando, Teófilo 72
High Council of Statistics (*Consiglio
　　Superiore di Statistica*– CSS) 163,
　　167–8
Higher Council for Scientific Research
　　(*Consejo Superior de Investgaciones
　　Científicas* - CSIC) xviii, 79–82,
　　87–95, 97–8, 103, 105–6, 108,
　　111, 125
higher education 41, 67–9, 83, 142, 184,
　　204 *see also* education
Himmler, Heinrich 15
Hinojosa, Eduardo 35
historiography xvii, 10, 66, 79, 8
holy scriptures 64–5
Houssay, Bernardo 180–1, 187, 190, 202
humanism 64
humanities 67, 91–2, 179, 183, 207

Ibañez Martín, José 85, 87–9, 91, 94–8
Illia, Arturo 189, 202, 214
imperial past 66
imperial science xviii, 10, 14, 79, 81–2,
　　87–9, 96–7
Inquisition 30, 64–5
intellectual crisis 60
intellectual isolation 16, 31, 38, 60, 64, 81,
　　109, 131, 203, 214
Inter-Ministerial Commission for
　　Insufficient Raw Material and for
　　Substitutes (CISS) 147, 155
internal exile 73, 106, 112, 208
International Atomic Energy Agency
　　(IAEA) 132
International Committee on Weights and
　　Measures (CIPM) 132
international community 70, 75
international economic crisis 145
International Monetary Fund 126
International Union of Pure And Applied
　　Chemistry 67
International Union of Pure and Applied
　　Physics (IUPAP) 132

Institute for Biological Studies (*Instituto de Investigaciones Biol*ógicas) 60
Institute of Clinical and Medical Research 111
Institute of Molecular Biology (*Instituto de Biología Molecular*) 75
Institute of Neurosurgery 114
Institute of Pedagogy (CSIC) 92
Institute of Spain (*Instituto de España*) 86, 90
Institute of Theology (CSIC) 91
Italian school of statistics 163
Italian Society for the Progress of Sciences (*Società Italiana per il Progresso delle Scienze* – SIPS) 167
Italian Society of Statistics' (*Società Italiana di Statistica* – SIS) 159, 161, 167–73, 175
Juan de la Cierva Centre 81, 91
Justo, Agustín P. 14, 181–2, 190

Kirchner, Néstor 182
Krause, Karl Ch. Fr. 31–2
    Krausism 31–4, 38
    Krausists 31–4, 36–8, 40, 44

Laboratory of Biological Chemistry (*Laboratorio de Química Biol*ógica –JAE) 67
Laboratory of Physiology (*Laboratorio de Fisiología*-JAE) 67
Lanusse, Alejandro A. 192, 204
Largo Caballero, Francisco 70
left wing 69, 187, 207
Leganés State Asylum 114
Levi, Mario Giacomo 154
liberal foreign trade policy 145
Liberal government 36, 38–9, 43, 49, 66
liberalism 2, 5, 14–15, 17, 21, 29, 31–3, 36, 38–9, 41–5, 49–51, 63, 66, 83–4, 95, 104, 106, 115, 181, 183, 202, 205
Lloret, Antoni 130, 135
London 7, 59, 74, 132, 170
Lora Tamayo, Manuel 86–7, 90

Lorente de No, Fernando 67
Lugones, Leopoldo 184

Macias, Ricardo 35
Magrini, Giovanni 143–4
Mantegazza, Paolo 180
Marañón, Gregorio 31, 44
MarcelinoMenéndez Pelayo Centre (CSIC) 91
Marconi, Guglielmo 143, 146, 152
Martínez de Perón, María Estela (Isabel) 191, 192, 207
Marxism 5, 71, 207
Mathematical Laboratory and Seminar 67, 70, 73
mathematics 7, 30–1, 42, 52, 59, 67, 69–70, 73, 127, 133, 142, 161–7, 169, 173, 183–4, 190, 193, 202, 212–13
Maura, Antonio 42, 62
mechanisation 64
Menem, Carlos 194–5
Menéndez Pelayo, Marcelino 31, 40, 91
Menéndez Pidal, Ramón 40, 48, 52–3, 68
Merton, Robert 2, 5, 28
Mexico 59, 74–5, 105, 106, 112, 114
Milia, Fernando 191
military coup 61, 67, 181, 186–7, 193, 202, 213–14
military plot 59, 69
Milstein, César 188
Ministry of Colonies 151,
Ministry of Public Education 152
Ministry of Public Instruction 50, 66,
Ministry of Public Works 66
Ministry of the National Education 152
Ministry of War 70, 142
moderate Republicans 70
Modern Age 64
modernisation 29, 32–3, 35, 38, 48, 53, 61, 65, 85, 104
Moles, Enrique 53, 59, 67–8, 70–2, 74,
Molinari, Henry 14, 19, 151
monarchy 33, 67
Montefinale, Tito 155
morality 27, 33, 35, 41, 48, 105

Mosconi, Enrique 182–3, 185
Museum of Anthropology 66
Museum of Natural Sciences 66
Mussolini, Benito 4, 14, 17, 141, 143–53,
    155, 163
Mutis, Celestino 64–5

National Academy of Sciences 180
National Action (*Acción Nacional*) 62
National Agency for Atomic Energy 14,
    185, 187, 191
National Cancer Institute 67
National-Catholicism 81–4, 87–91, 94–8,
    124
National Centre for Scientific Research
    (*Centre National de la Recherche
    Scientifique*) 74
National Council for Physics (*Consejo
    Nacional de Física,* CNF) 132
National Council for Research (*Consiglio
    Nazionale delle Ricerche* – CNR)
    xviii, 11, 13–14, 17, 19, 86,
    141–53, 155–6
National Council of Scientific and
    Technological Research
    (CONICET) 187, 189–90, 195,
    207, 213–14
National Foundation for Scientific Research
    and Trials of Reform *(Fundación
    Nacional para Investigaciones
    Científicas y Ensayos de Reformas* –
    FNICER) 11, 18, 38–9, 69, 86, 88,
    92, 107
National Geophysical Service 153, 156
national identity 66
National Institute for Medical Sciences
    105
National Institute for Physical -Natural
    Sciences (JAE) 52
National Institute for the Applications of
    Calculation 155
National Institute of Farming Technology
    187

National Institute of Food Hygiene
    (*Instituto Nacional de Higiene de la
    Alimentación*) 70, 110
National Institute of Industrial Technology
    187
National Institute of Oncology 67
national laboratories 7, 144, 151, 156
National Meteorological Service 180
National Pedagogic Museum 92
National Physics and Chemistry Institute
    (*Instituto Nacional de Física y
    Química*) -JAE) 52–3, 63, 67–8,
    70, 90
National School of Childcare 107
National Secondary School of Buenos Aires
    (*Colegio Nacional de Buenos Aires*)
    180, 183
nationalism 7, 11, 14, 16, 18–19, 21,
    60–62, 80, 86, 103, 113–14, 142,
    146, 154, 170, 174, 182–4, 194,
    201, 207
national science xiii, 13, 19
navy 124, 137, 185, 192
Nazi 4, 15, 19, 172, 175
Negrín López, Juan 44, 67, 72–3
network of influence 48, 51
Nerve Centre Physiology and Anatomy
    Laboratory (JAE) 52
Nervous System Histopathology
    Laboratory (JAE) 52
Neurohistopathology 104, 108
Neurolathyrism xviii, 114
Neurology 105, 109–14
Neurosurgery 105, 114
New Spain 63, 69, 90, 92–3
New Regime 13, 63, 68, 71–2, 84–6, 96
Niacin Deficiency 70
Night of the Long Sticks (Noche de los
    Bastones Largos) 189–91, 204, 208
Nobel Prize xv, 9, 13, 20, 34, 60, 66, 73–5,
    104, 115, 180, 188
Nuclear Energy Board (*Junta de Energía
    Nuclear,* JEN) 14, 124–5, 129, 135

Obermaier, Hugo 71

Ochoa, Severo 53, 71, 75
Onganía, Juan Carlos 189, 191, 204
Opus Dei 87, 108
Ortega y Gasset, José 53, 60, 71
orthodoxy xviii, 2, 28, 64, 160, 170

paediatrician 60, 85, 107
Palacios, Julio 46, 60, 68, 73, 86–7, 90
Palaeontology and Prehistory Research
        Commission (JAE) 52
pandemic disease 70
panentheism 32
Paresthesia-Causalgia Syndrome 113
Parravano, Nicola 145
pedagogy 33, 45, 60, 63, 66, 92–3
Pemartín, José 83
Pérez-Vitoria, Augusto 59, 61, 75
Permanent Delegation of the Republican
        Parliament 73
Perón, Juan Domingo 5, 8, 12, 185–7,
        191–2, 201–2, 204–5, 213
    Peronism 4–5, 8, 12, 185–9, 191–2,
        195, 201–5, 207
philosophy 8, 31–2, 34, 37, 60, 66, 89, 124,
        192, 200, 206, 208, 212–14
physics xviii, 6–8, 13, 15, 20, 42, 52–3,
        59–60, 63, 67–71, 73–5, 86, 90,
        121–37, 142, 156, 192, 212
Physics Research Laboratory 42, 52
physiology 52, 60, 67, 71, 73, 75, 108, 111,
        180
Pí Calleja, Pedro 73
Picatoste, Felipe 31
Pirosky, Ignacio 188
Polanyi, Michael 2, 5, 28
policy of colonial conquest 145
poliomyelitis 109
Popular Front 69, 110
positive mentality 34
Prieto, Indalecio 73
Primo de Rivera, Miguel 43–4, 67–8, 107
probability 160–7, 169–71, 173, 175
Protestantism 64, 83
pseudoscience 2–4
public instruction 30, 38, 42, 50, 62 3, 70

Pulqui 185–6
pure science 39, 69, 92, 142
purge xvii, xviii, 11–14, 16–17, 19, 21, 59,
        61–3, 71–2, 80, 85–7, 90, 92, 105,
        203, 207

Quartino, Bernabé 195

racial laws 12, 16, 18, 154, 168
Ramón y Cajal, Santiago 11, 17, 31, 34–7,
        40, 48, 52, 60, 66–7, 91, 104–5,
        111
Raimundo Lulio Centre (CSIC) 91
Recasens, Luis 71
Re-Christianization 182, 184
reconciliation 59, 61, 75–6
Red Revolution 63
Reformists 63, 66, 181, 187–8, 200–203
refugees 73–4
regeneration 34–8, 48, 60, 65–6, 83, 96
regime xvi, xvii, xviii, 1–7, 9–16, 19, 21,
        29, 33, 59, 62–3, 68, 71–3, 75, 79,
        81–7, 95–9, 103, 106–10, 112–13,
        115, 122–5, 130, 144, 152, 159,
        168–9, 171–4, 186–7, 189, 201,
        204, 207
religion 10, 30, 35, 82, 88–9, 90–2, 94,
        97–8, 201
Republican 33, 41, 44, 60, 62–3, 66, 68–74,
        87, 103–4, 107, 110, 112–15, 124,
        132
    Republican Administration 72
    Republican refugees 74
Revilla, Manuel 30
Rey Pastor, Julio 31, 42, 59, 67, 73, 202,
        213
Richter, Ronald 5, 185
right wing 69, 73, 182–4, 187–8, 190, 207,
        213
Rivera, Julian 35
Rockefeller Foundation 53, 67–8, 109, 111
Rodríguez Carracido, José 31, 34–6, 40
Rodríguez Lafora, Gonzalo 53, 71, 105–7,
        112
Rosas, Juan Manuel de 179

Royal Botanical Garden 66
ruralist perspective 145

Sacheri, Carlos Alberto 190
Sacristán, José M. 53, 71
Sadosky, Manuel 194, 203
Sáinz Rodríguez, Pedro 86, 90, 95
Salmerón, Nicolás 31–2, 34
sampling 159, 163, 165–7, 175
Sánchez Albornoz, Claudio 60, 71
Sánchez Guerra, José 62
Sánchez, Raúl 192
Santaló, Miguel A. 73
Santiago Ramón y Cajal Centre 91, 105,
    111
Sanz del Río, Julián 31
Sarmiento, Domingo Faustino 179–81
Savio, Manuel 14, 182–3
scholarships 45, 66, 75
School of Agronomical Engineering of
    Madrid 107–8
School of Criminology (*Escuela de
    Criminología*) 63
science
    science policy xiv, xv, xvi, xvii, xix, 1–2,
        7, 11, 18, 20–21, 27, 29, 36–7, 42,
        85–6, 95, 104, 122, 135, 199
    scientific beliefs 61
    scientific communities xvi, xviii, 2–6,
        10–18, 21, 28, 49, 60–1, 80, 82,
        84–5, 92–3, 98–9, 141, 156, 185,
        189, 195
    scientific development xiii, 1–2, 7–8,
        10, 18, 20–21, 28–9, 52, 84, 96, 98,
        179, 181, 183, 191
    scientific education 18, 31, 36, 203
    scientific institutions xiv, xvi, 4, 7, 11,
        37, 60, 104, 193
    scientific integrity 2
    scientific knowledge xvii, 20, 70
    scientific missions (Italy) 151
    scientific productivity 200
    scientific research 2, 7–8, 17–18, 21,
        27, 31, 33, 36–9, 41, 51–3, 64, 69,

    79–80, 86–7, 90, 95, 103, 126, 131,
        150–51, 153, 169, 203, 211
    scientific-technical nationalism 146,
        154
    Scientific Revolution 64, 80–1, 87
Second Republic 18, 39, 41, 44, 60–1,
    103–4, 107
Second World War xvi, xviii, 3, 7, 15, 28,
    62, 81, 83, 114, 141, 150, 155, 159,
    172, 183, 185
secrecy 3, 5, 62, 84, 148–9, 188, 193
secularism 66, 104, 115, 180
Sepich, Juan 183
Service for the Evacuation of Spanish
    Refugees (*Servicio de Evacuación de
    Refugiados Españoles*– SERE) 73–4
Simarro, Luis 40
social contract for science xvii, 2–4, 7–8,
    15, 27–9, 34, 37, 39, 41–2, 48–9,
    52–3, 80
Socialism 13, 63, 70, 83
Socialist Party 12, 73
Society of Demography and Statistics
    (*Società italiana di demografia e
    statistica* – SIDS) 168–70, 173, 175
Spanish Action 62
Spanish border 59, 64
Spanish Civil War *see* Civil War
Spanish Empire 65, 136
Spanish Golden Age 62, 68, 97
Spanish-Mexican Academy 74
Spanish Second Republic *see* Second
    Republic
Spanish Second Republic Provisional
    Government 68
Spanish Society for Physics and Chemistry
    90
spiritual 32, 83, 85, 89, 95–7, 124
Students Residence (*Residencia de
    Estudiantes*) 52, 63, 70, 80, 92
Summer Universities 64
Supreme Commission for Defence
    (*Commissione Suprema di Difesa* –
    CSD) 148, 152
Suñer, Enrique 60, 85, 107

Tagüeña, Manuel 59
Tank, Kurt 185
Teachers Training College (*Escuela Superior del Magisterio*) 63
Technical Committee to Aid Spanish Refugees 74
technocracy 98
Tello Muñoz, Jorge F. 67, 71, 106, 108
tension between faith and reason xviii, 81, 87–8
Terradas, Esteban 59, 73
theory of the two demons 199
Third Spain 60, 71
Torres, Leonardo Quevedo 40, 60
tradition 6, 11, 16, 19, 21, 35, 61–2, 64–6, 69, 73, 75, 80, 87, 90, 94–7, 103, 115, 124, 175, 179, 184, 190
    traditionalist 30–1, 69, 84, 88, 91

Unamuno, Miguel de 64,
Union of Spanish University Professors Abroad (*Unión de Profesores Universitarios Españoles en el Extranjero*) 73

United Nations 126
unity of sciences 89, 91–2, 94, 98
University of Buenos Aires 179, 180, 186–92, 201–6, 213
University of the Catacombs 208
university reform 181, 192, 200, 212
Uriburu, José Félix 181–2

Vélez, Oscar G. 191
Vera, Francisco 73
Videla, Jorge Rafael 183, 207, 209
Volta, Alessandro 7
Volterra, Vito 7, 142–3, 212

war effort 9, 19, 70, 155, 160, 163, 172, 175
Workers' High Schools (*Institutos para Obreros*) 70

Yrigoyen, Hipólito 181

Zubirán, Salvador 74
Zubiri, Xavier 71